Northern Colorado Plateau

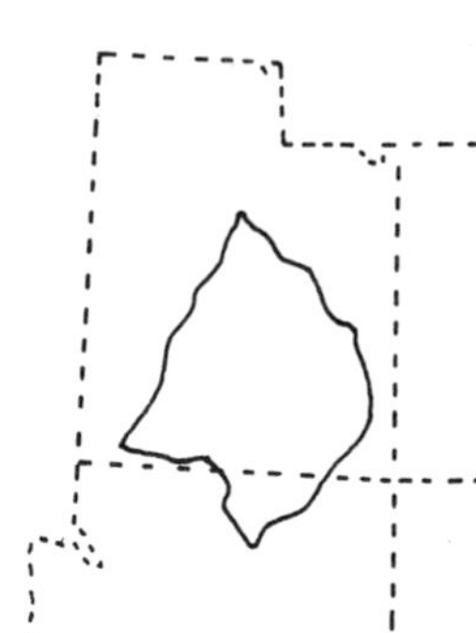

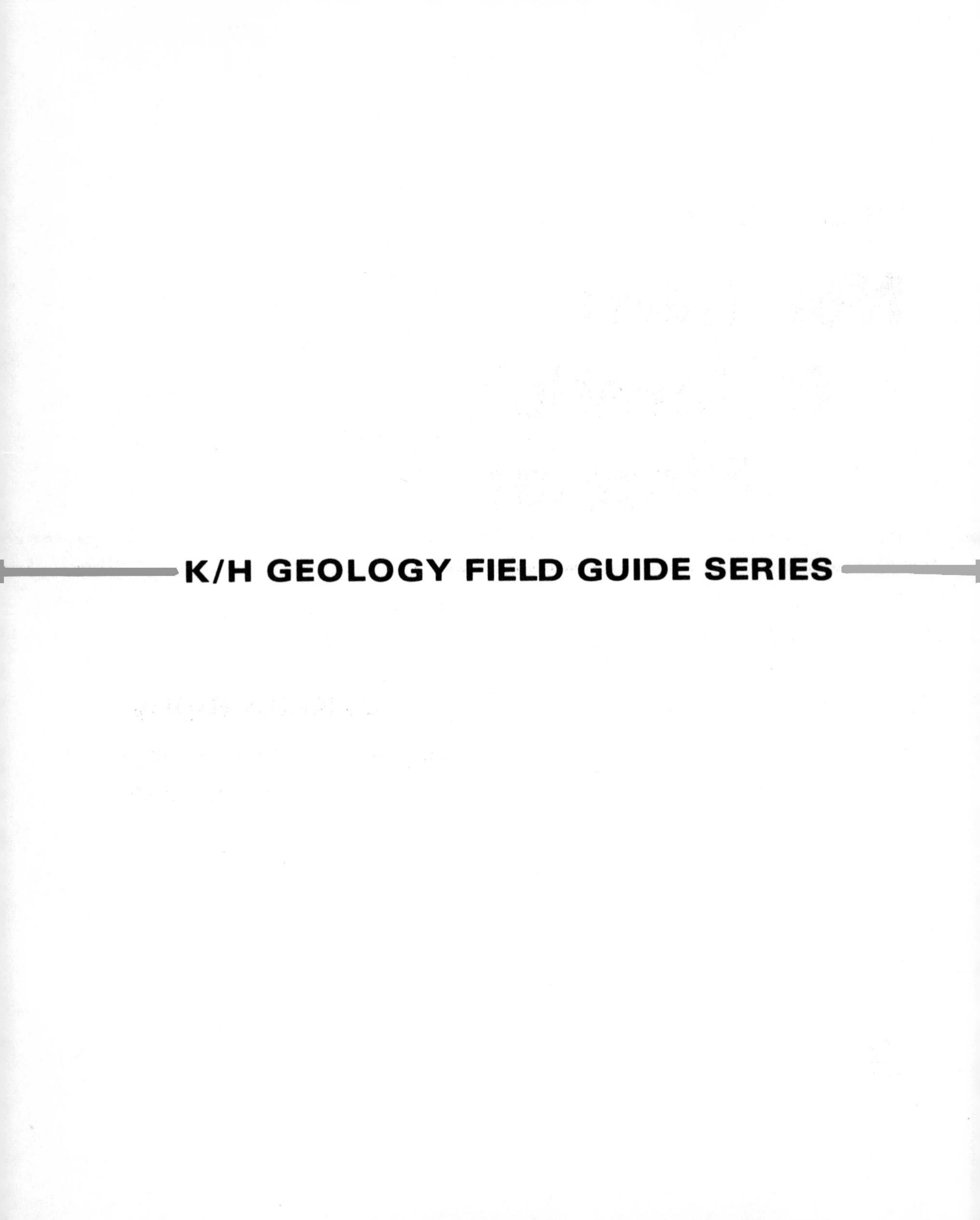
K/H GEOLOGY FIELD GUIDE SERIES

FIELD GUIDE

Northern Colorado Plateau

J. Keith Rigby
Brigham Young University
Provo, Utah

KENDALL/HUNT PUBLISHING COMPANY
Dubuque, Iowa

Library of Congress Catalog Card Number: 75—32218

ISBN 0—8403—1313—6

Printed in the United States of America

CONTENTS

INTRODUCTION

The Colorado Plateau in the Four Corners area in western North America is a scenic area and is an easy area to study geology because the rocks are colorful and unusually well exposed. The combination of high relief and an arid climate has produced limited plant cover. As a consequence soil cover is removed and transported to the lowlands or to the sea by fast-flowing streams. The Colorado Plateau can be subdivided into six major regions or sections on the basis of features of relief and kinds of geology (fig. I.1). It is rimmed on the northwest by the High Plateaus Section, on the north by the Uinta Basin section, and on the east it is flanked by the Colorado Rocky Mountains. The central section, of much concern to us in the present guidebook, forms relatively low country in the Canyonlands section which is bissected by the Colorado River and its tributaries. To the southeast the Navajo section covers much of the country immediately south of the Four Corners area in northern Arizona and New Mexico. In the southwest is the Grand Canyon Section, an area made famous by the scenery and tremendous exposures in the great canyon of the Colorado River, where the river has cut through between the Kaibab and Coconino plateaus. Far to the south and southeast, beyond the guidebook coverage, is the Datil section of the Colorado Plateau, an area dominated by volcanic rocks. It is primarily with the High Plateau section, the Canyonlands section, and parts of the Grand Canyon and Navajo sections that the present guidebook is concerned.

The southern margin of the Colorado Plateau (fig. I.1) is drawn at the Mogollon Rim, an erosional escarpment at the edge of the flat uplands. Here Paleozoic and Mesozoic rocks of the escarpment look down over the older folded rocks, as old as Precambrian, in central and southern Arizona and in parts of New Mexico. The west side of the Colorado Plateau is drawn along major faults of the eastern Great Basin and the Basin and Range. North-

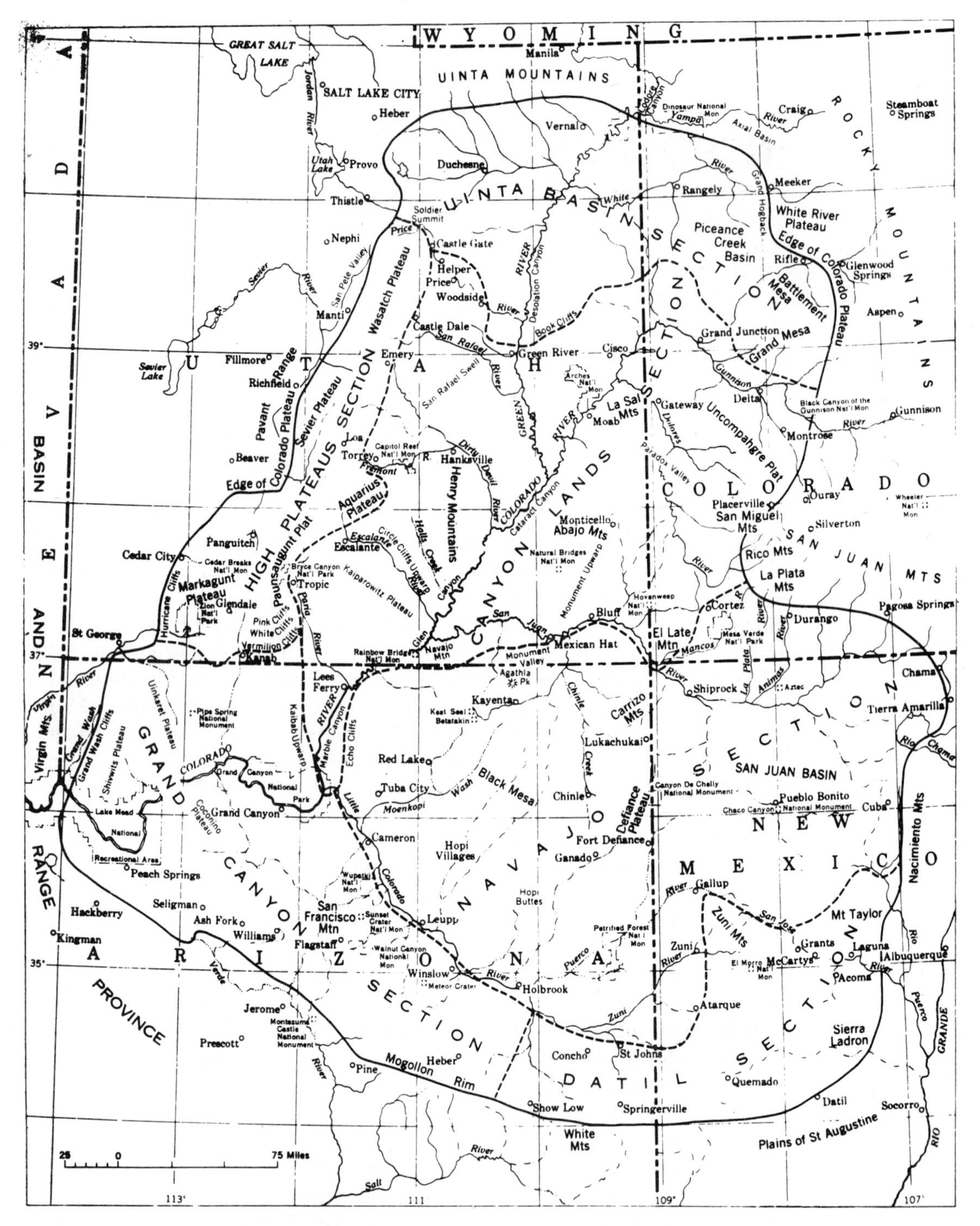

Figure I.1. Index map to geographic features of the Colorado Plateau and the immediately surrounding area (Modified from C.B. Hunt, 1956, Cenozoic Geology of the Colorado Plateau: U.S. Geological Survey Professional Paper 279).

western and northern limits of the plateau through central and northern Utah are drawn east of the folded and faulted Wasatch Mountains and Pavant Range and south of the upfolded Uinta Mountains. The eastern margin of the plateau is along the western edge of various ranges of the Colorado Rocky Mountains.

The Colorado Plateau is a region dominated by plateaus and tablelands, by mesas and buttes, punctuated here and there by volcanic peaks or by igneous-cored mountains that rise above the dominantly deeply-inscribed plateau country. Major volcanic centers, like those of the San Juan Mountains, the Carrizo Mountains, San Francisco Peaks, Marysvale and Fish Lake volcanic areas occur around the margin of the outer edges of the plateau providence (fig. I.2). Laccolithic intrusions, first described in the Henry Mountains of the plateau by G.K. Gilbert, core the LaSal Mountains, the Abajo Mountains, the Henry Mountains and some of the other peaks of the plateau.

The High Plateaus Section of the Colorado Plateau, in general, occurs east of U.S. Highway 91 and west of the large San Rafael Swell and Circle Cliff Uplift. The high plateaus extend southward from U.S. Highway 50-6 near Provo and Price to Bryce Canyon and Zion Canyon National Parks, near the Arizona-Utah border. The High Plateaus are made up of a thick series of elevated, somewhat faulted, Tertiary rocks and are capped with peaks 10,000 to 12,000 feet high. The eastern escarpment is abrupt from Price to the south and is part of the Book Cliffs. These cliffs along the eastern edge of the Wasatch Plateau, Aquarius Plateau, and Paunsaugunt Plateau mark the boundary between higher country of Tertiary and Cretaceous rocks, on the west, and the generally lower country of Triassic and Jurassic rocks of the Canyonlands section, to the east. High relief in the High Plateaus is more a matter of valleys and canyons being carved into the flat residual uplands than it is to these areas being raised above the level of the country around.

The western margin of the High Plateaus coincides approximately with the main front of major Sevier folding which took place during the Cretaceous and early Tertiary. Much of the sedimentary rocks of the northern High Plateaus record the coarse debris shed eastward off those ancient mountains into the low country along the piedmont or base of the mountains, prior to the uplift of the rocks to their present elevation. These coarse rocks are particularly well shown in the north in the Wasatch, Sevier, and Gunnison Plateaus. Eroded roots of these ancient mountains are well exposed west of the plateaus in the Wasatch Mountains and in some of the other folded and intricately faulted mountain ranges along the general route of U.S. Highway 91. Toward the south the Markagunt Plateau and Paunsaugunt Plateau similarly show Tertiary and older sedimentary formations but these rocks were deposited at some distance from their source and are distinctly finer grained.

The intervening Sevier, Fish Lake, and Awapa Plateaus in the central part of the High Plateaus Section are carved in high volcanic fields associated with igneous intrusions and lava fields that centered in the Marysvale region of the plateau.

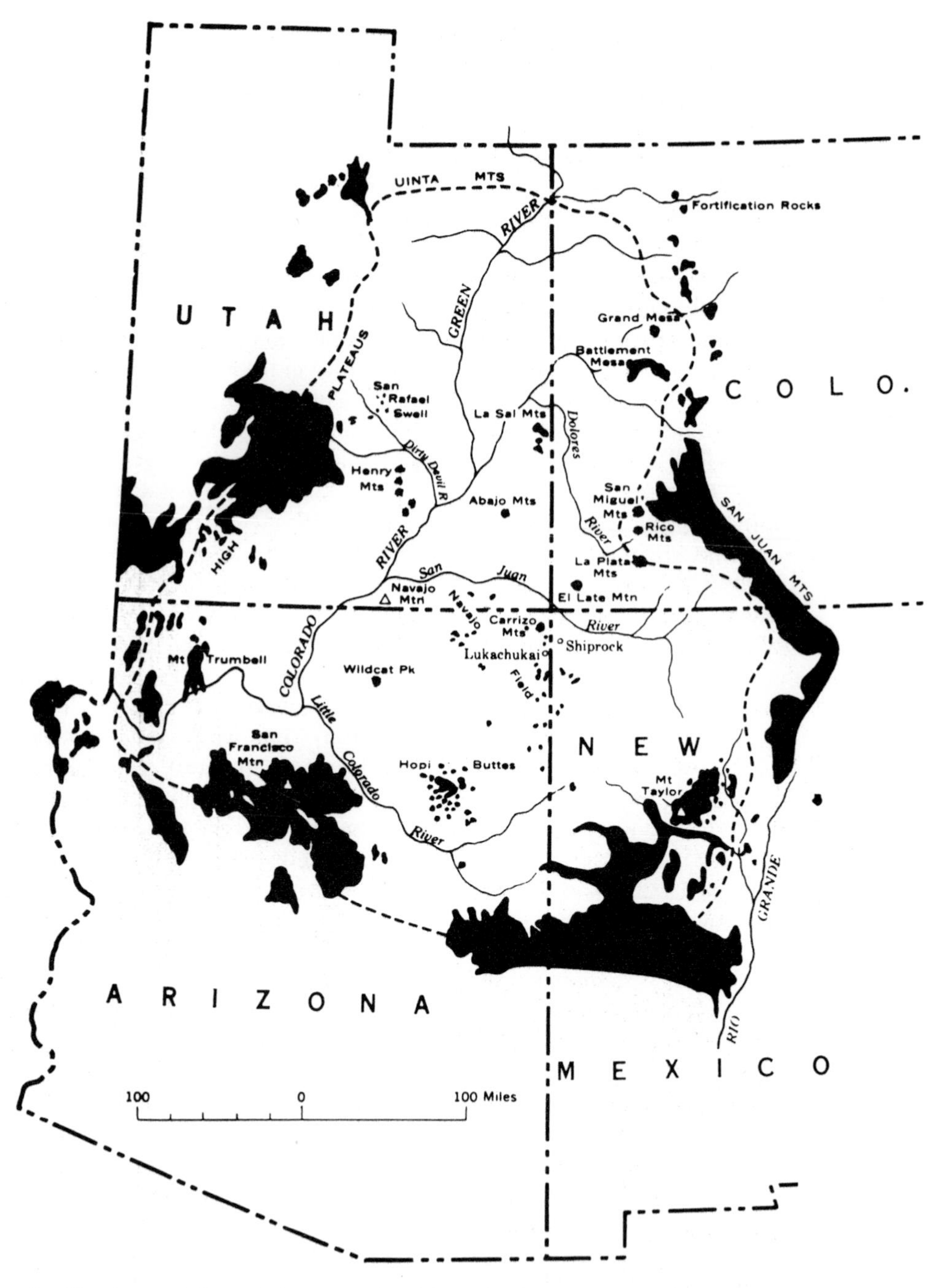

Figure I.2. Sketch map of Cenozoic igneous rocks of the Colorado Plateau showing the major volcanic fields around the edges of the plateau. Laccolithic Mountains and other small intrusions are more common in the central part of the plateau (From C.B. Hunt, 1956, Cenozoic Geology of the Colorado Plateau: U.S. Geological Survey Professional Paper 279).

The High Plateaus are divided into three north-south trending blocks by two major faults, the Sevier Fault which roughly parallels U.S. Highway 89, and the Paunsaugunt Fault which separates plateaus on the east, such as the Aquarius Plateau from the Paunsaugunt Plateau on the west. The Hurricane Fault, or its *en echelon* somewhat north-northeasterly continuation, the Wasatch Fault, blocks out the western margin of the Markagunt Plateau and the Tushar Mountains, Pavant Range, Valley Mountains and Gunnison Plateau. The Sevier Fault also continues along the eastern border of these ranges, northward along the western base of the Sevier Plateau and the Wasatch Plateau. North of Salina the western margin of the Wasatch Plateau grades into a major monocline, another feature common as a structural element in the Colorado Plateau in the broad sense. The Sevier Fault, where it grades northward into the Wasatch Monocline, has displacement of only a few hundred feet but along the monocline there is structural relief of as much as 7,000 feet where the plateau is uplifted above the equivalent rocks in the valley.

The south-facing escarpment of the High Plateaus is sometimes called the Grand Staircase because of the steplike erosion of the alternating resistant and nonresistant nearly horizontal Mesozoic and Cenozoic beds as they rise above the Canyonlands and the Grand Canyon sections. The Chocolate Cliff, Red Cliffs, White Cliffs, Gray Cliffs and Pink Cliffs step upward from the section margin to the top of the Paunsaugunt Plateau.

The Uinta Basin Section, on the north, is principally carved in Tertiary rocks. These rocks occur in a broad downwarp into which major accumulations of Tertiary sediments were deposited from uplifts in the west and north. The Uinta Basin Section is the principal oil shale area within the Colorado Plateau, where the Eocene Green River Formation is still well preserved. The Book Cliffs and Roan Cliffs form an effective barrier around the southern edge of the Uinta Basin Section. This high escarpment clearly delimits the high upland of the East and West Tavaputs Plateaus, on the north, from the Canyonlands area along the guidebook route on the south. The broad, wide, easily eroded belt of gray Cretaceous shale forms the steep badlands at the base of the escarpment. In general, the boundary between the High Plateau Section and the Uinta Basin is drawn through the low passes along U.S. Highway 50 and 6.

The Uinta Basin is a topographic basin drained by the through-flowing Green River which cut Desolation Canyon across the boundary of the section. Average elevation of the Uinta Basin section is over 5,000 feet even though it is structurally one of the lowest parts of the entire Colorado Plateau complex (fig. I.3). The whole plateau appears somewhat like a broad saucer, tilted toward the northeast, with old rocks exposed around the rim but with youngest rocks exposed here in the Uinta Basin Section. Routes outlined in the guidebook skirt around only the southern margin of the basin but the southern edge shows very well in the vicinity of the Book Cliffs.

The Uinta Basin Section is a broad asymmetric syncline with very gentle dips along the south and the eastern margin but with steeply upturned Mesozoic and Paleozoic beds exposed along the western and the northern margin. Unlike the High Plateaus, where the separate plateaus are blocked out by active faults, the southern margin of the Uinta Basin Section is merely an erosional escarpment on gentle beds dipping off the domes and anticlines to the south into the Uinta Basin, on the north, or Piceance Creek Basin, to the northeast.

The Grand Canyon section in the southernmost part of Utah and in northern Arizona is centered west of the broad flatlands of the Navajo Section in the Four Corners region and south of the High Plateaus. Along the guidebook route we cross through the northernmost margins of the Grand Canyon Section only in the area around Kanab, and here Triassic rocks are well exposed around the base of the High Plateaus. It is on these relatively easily eroded formations that the highway has been constructed and it is these subsequent valleys that form the boundary at the north edge of the Grand Canyon Section and the south edge of the High Plateaus. The Grand Canyon Section and much of the Navajo Section are more extensively covered in a companion guidebook on the Southern Colorado Plateau and Four Corners region. The Grand Canyon Section is structurally the highest part of the Colorado Plateau and marks the uplifted southern edge of the broad "saucer." Here Precambrian rocks are exposed in the deeply entrenched Colorado River Canyon in Grand Canyon National Park. Rocks dip toward the north and northeast in much of the section and it is the subsequent valleys along the northeastern part of the block that are along our guidebook route.

The Navajo Section of the Colorado Plateau is a somewhat less clearly defined section and extends in general south and east of the intricately carved Canyonlands along the Colorado River drainage, south of the Four Corners region of the plateau. It is north of the volcanic fields of Datil section and east of the major monoclines of the Grand Canyon Section. We cross part of the Navajo Section from near Page, Arizona and Tuba City northeastward through Kayenta into Monument Valley around the northern edge of Black Mesa and the uplifts west of the San Juan Basin.

On our route we traverse principally Triassic and Jurassic rocks in the northwestern part of the Navajo Section. Here younger rocks are preserved only in broad basins, such as that along Black Mesa and eastward in the San Juan Basin.

The Navajo Section has nearly as high an average elevation as the Canyonlands Section to the north but is less intricately carved. Even the main streams are less deeply entrenched. In general the whole region is characterized by gently warped to nearly flat lying rocks. Only a few sharply folded monoclines and broad regional upwarps affect the geological structure and exposures of the beds so that buttes and mesas are more common than the steep hogbacks that are observed in some of the other more sharply folded rocks.

Lava flows, dikes, sills, and volcanic necks are encountered with more commonness in the Navajo Section (fig. I.2) than in other sections of the plateau where laccolithic and stocklike intrusions are more common. For example, intrusions create prominent features near Shiprock, New Mexico area or in Agathale's Needle near Kayenta, Arizona. Agathale's Needle for example rises 1,200 feet above its intruded base on the Chinle Shale and is a striking spirelike feature along the route.

The Canyonlands Section is the other major section in the northern plateau which is visited on our guide. It is rimmed by highland sections but is carved into spectacular scenery by entrenchment of the Colorado River and Green River and their tributaries. Pennsylvanian and Permian rocks are exposed in some of the broad uplifts, such as the Monument Valley uplift, along the deeply entrenched canyons, but in general the entire upland surface of the Canyonlands Section is carved in Triassic and Jurassic rocks and many of these are brightly colored and add to the scenery for which the Canyonlands area is famous. It is principally within the area of immediate drainage in the Colorado River that canyons are prominently developed in the region. Broad flat divides are common and relief is relatively slight near where the Canyonlands Section joins with the High Plateaus or the Uinta Basin Sections or to the Navajo Section to the southeast. One of the most prominent of these broad flatlands is the Great Sage Plain in the southeastern part of Utah and the Four Corners area, north of the San Juan River and east of Monticello.

Geologic structure of the northern part of the Canyonlands Section is dominated by the large domal San Rafael Swell (fig. I.3). The southeastern part of the section is over the broad Monument Valley upwarp and that to the southwest is over the Circle Cliffs upwarp. These upwarps surround broad basins, like the Henry Mountains Basin, and are commonly bounded by moderately steep monoclinal flexures at least on one or the other limbs.

Another characteristic structural feature of the Canyonlands Section is the northwest-southeast trending salt anticlines of the LaSal Mountains-Moab area. Here long cigar-shaped accumulations of salt have uparched the overlying Mesozoic beds. With subsequent removal of the salt, "keystones" have collapsed. These collapsed "keystones" form broad elongate valleys now rimmed by uptilted edges along their flanks. Some of the strongly jointed keystones have produced some spectacular scenery in the Arches National Park and Spanish Valley areas along the guidebook route.

Major mountain peaks like those in the Henry Mountains, the LaSal Mountains, the Abajo Mountains rise above the route and are characteristic of some of the mountains cored by laccolithic igneous intrusions. Indeed it was in the Henry Mountains that G.K. Gilbert first postulated the existence of mushroom-shaped intrusions, which he termed laccoliths.

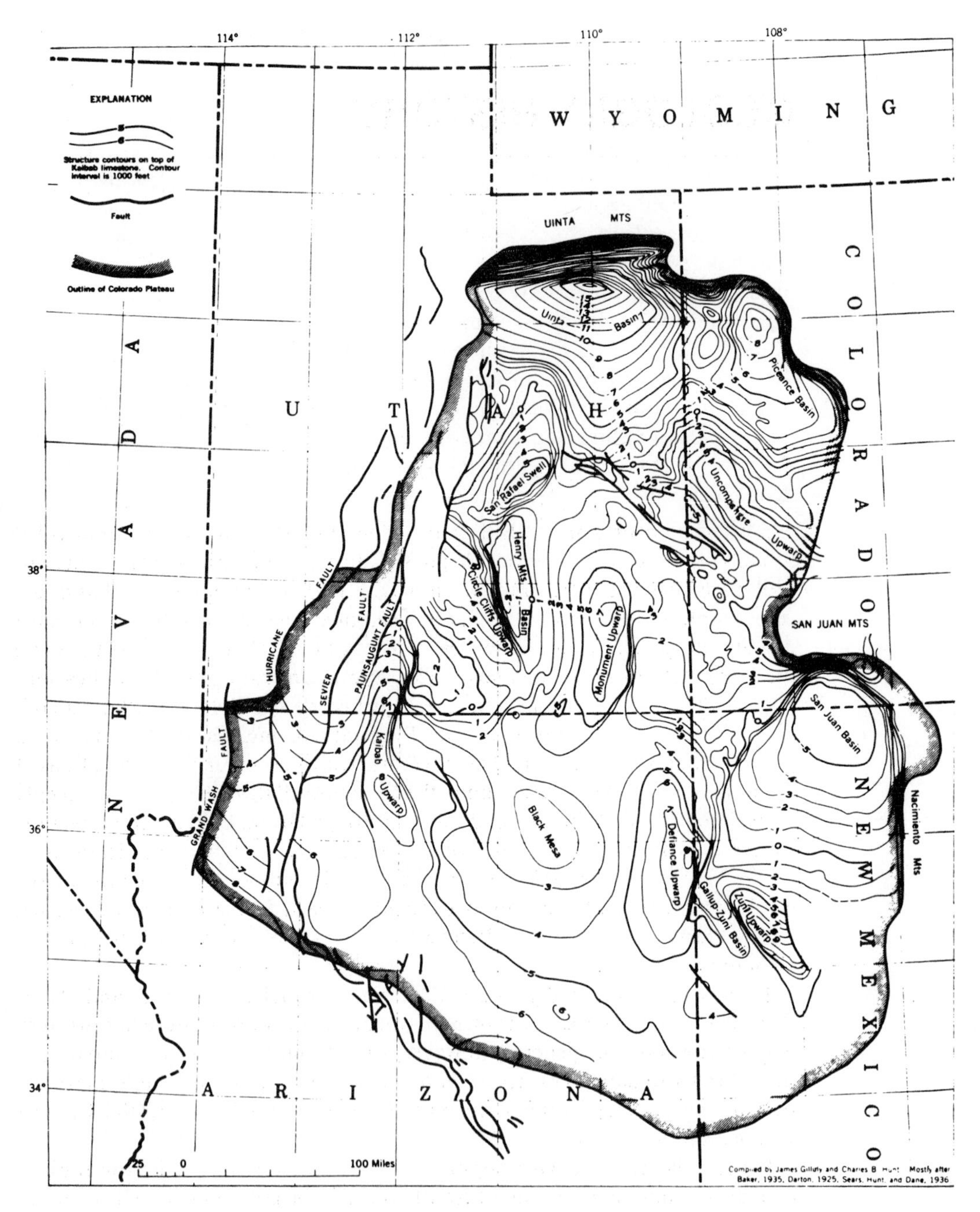

Figure I.3. Structural contour map of the Colorado Plateau showing the major uplifts and basins as they are expressed by elevations of the top of the Kaibab Limestone or its equivalents (From C.B. Hunt, 1956, Cenozoic Geology of the Colorado Plateau: U.S. Geological Survey Professional Paper 279).

GEOLOGIC HISTORY

Geologic history of Utah and the Colorado Plateau can be subdivided broadly into six major periods or phases, during which the geologic behavior of the area has had different patterns (fig. I.4). The first phase was that of accumulation of thick deposits of sandstone, limestone, shale in the western part of the state generally along that region west of U.S. Highway 91. At the same time thin sections of sedimentary rocks were accumulating in the southeastern part of the state, in general that area covered by the Colorado Plateau (fig. I.5A). Phase I lasted through Late Precambrian into Devonian time, or a period of approximately 300,000,000 years. Rocks accumulated during Phase I are not seen in general throughout the Colorado Plateau except in the deeply entrenched gorge of the Colorado River through Grand Canyon and along the southern escarpment at the edge of the plateau in central Arizona. In the area covered by the guidebook, however, the thick sedimentary sequence of the northwestern part of the state is exposed, in part, in the folded and faulted ranges through which U.S. Highway 91 cuts from Provo south toward St. George.

Thick sections of Precambrian, Cambrian, Ordovician, Silurian and Devonian rocks accumulated in the miogeosyncline in the northwestern and western part of the state. In general, however, only Cambrian, thin Devonian, and some Mississippian rocks accumulated along the craton, the staple platform part of the eastern southeastern part of the state during the initial Phase I sedimentation.

Phase II lasted from Mississippian time into early Triassic time for approximately 150 million years, and here the general pattern is dominated by major northwest-southeast trending basins and accompanying uplifts (fig. I.5B). During this time thick sequences of rock accumulated in the Oquirrh Basin in a region near Nephi, Provo, and to the northwest. At about the same time thick

Era	Period		Epoch	Millions of Years		Author, Type Locality, and Reference Areas in North America	Diastrophic Events
Cenozoic	Quaternary		Recent	Last 10,000 Years		Jules Desnoyers proposed the Quaternary in 1829 for young deposits in France.	
		Neogene	Pleistocene	1-2.5	1-2.5		
	Tertiary	Neogene	Pliocene	6	7	Giovanni Arduino defined Tertiary rocks in 1760 in Italy as separate from the Primary and Secondary. Epochs were proposed by Charles Lyell in 1883 from type sections in France. Tertiary rocks are extensive in the High Plains and Coastal Plains in North America.	
		Neogene	Miocene	19	26		Coast Ranges Basin and Range
		Paleogene	Oligocene	12	38		
		Paleogene	Eocene	16	54		
		Paleogene	Paleocene	11	65		
Mesozoic	Cretaceous			71	136	d'Omalius d'Halloy proposed the term Cretaceous in 1822 for the chalk-bearing sandstones and shales around the Paris Basin in France. Cretaceous rocks are well developed in the Coastal Plains and western interior of Canada and the United States.	Laramide
	Jurassic			54	190	Alexander von Humbolt named the Jurassic from rocks in the Jura Mountains in northern Switzerland in 1795. Leopold von Buch proposed it as a system in 1839. Much scenery of the Colorado Plateau is carved into Jurassic rocks.	Cordilleran
	Triassic			35	225	Friedrich von Alberti introduced the term Triassic in 1834 for the three-fold subdivision of salt-bearing rocks in northern Germany. Triassic rocks are extensive in the Mountain West.	Palisades
Paleozoic	Permian			55	280	Roderick Impey Murchison named the Permian System from the province of Perm on the west flank of the Ural Mountains in Russia. Reef-associated exposures in West Texas and New Mexico are standards for North America.	Appalachian
	Pennsylvanian	Carboniferous		45	325	Henry Shaler Williams proposed the Pennsylvanian System in 1891 for the coal-bearing rocks of Pennsylvania, equivalent to the Upper Carboniferous of Britain.	Marathon Colo. Mtns.
	Mississippian	Carboniferous		20	345	Alexander Winchell proposed the Mississippian System in 1870 for the Lower Carboniferous rocks exposed in the Mississippi River Valley.	
	Devonian			50	395	Roderick Impey Murchison and Adam Sedgwick proposed the Devonian System for rocks in Devonshire, England. Devonian rocks are classically developed in New York and Pennsylvania.	Acadian, Antler, and Ellesmerian
	Silurian			35	430	Roderick Impey Murchison proposed the Silurian System in 1835 for exposures in southern Wales. Silurian rocks are well developed in the Niagara Gorge region of New York and Ontario.	
	Ordovician			70	500	Charles Lapworth named the Ordovician System in 1879 for a series of rocks involved in a controversy concerning the boundary of the Cambrian and Silurian Systems. Ordovician rocks are well developed in New York, Ohio, Utah, and Nevada.	Taconic
	Cambrian			70	570	Adam Sedgwick proposed the Cambrian System in 1835 for a series of shale and sandstone in northern Wales. Excellent exposures of Cambrian rocks occur in California, Utah, Alberta, British Columbia, and Wisconsin.	
	Precambrian			4,030		Precambrian rocks are exposed in the cores of continents. In North America the most extensive exposures of Precambrian rocks are in the Canadian Shield surrounding Hudson Bay. They are also exposed in cores of the Rocky Mountains and Appalachian Mountains. Grand Canyon of the Colorado River exposes classic sequences. Precambrian rocks are essentially unfossiliferous.	Numerous

Figure I.4. Geologic time scale (From Petersen, Rigby, and Hintze, Historical Geology of North America: Wm. C. Brown Co., Publishers).

sections of rocks also accumulated in the Paradox Basin, in the southeastern part of the state, in the Four Corners region in general. The Paradox Basin is flanked on the north by the Uncompahgre Uplift. It was the interaction of these two structural units during Late Paleozoic time that produced bedded rocks which ultimately have eroded to produce the spectacular scenery of the eastern part of the plateau. The bright red rocks, for example, seen in the vicinity of Moab or to the northeast of Moab represent iron-bearing sediments shed from the Uncompahgre Uplift southwestward into the Paradox Basin. These red beds represent nonmarine sediments that interfinger with marine limestone and shale in the interior of the basin. This interfingering sequence has produced some of the spectacular striped country of Canyonlands, Monument Valley, and of the general Lake Powell area. Phase II lasted from approximately 350,000,000 years ago to about 200,000,000 years ago and effects of the basins and accompanying uplifts, in controlling sediments, was strong until the beginning of Phase III in the Late Triassic.

Phase III encompasses most of Mesozoic and earliest Cenozoic time and, in general, was a time during which the major Sevier mountain building movement occurred in the miogeosynclinal area (fig. I.5C). The region generally northwest of U.S. Highway 89 and 91 was folded and faulted and shed detrital deposits into the plateau area and the southeastern part of the state. Miogeosynclinal rocks in western Utah were folded and faulted to produce rugged mountains from which mud, sand, and coarse gravel were spread eastward into the low lying country. The area of deposition was almost at sea level and covered much of the eastern and central parts of the plateau. The early part of Phase III was marked by repeated invasion and retreat, of marine waters from the north and from the south and development of broad thin sheets of marine deposits and accompanying marginal tidal flat deposits. The later part of Phase III is that during which the Sevier Orogeny produced a long mountain chain along the western part of North America and, more locally, affected the miogeosynclinal rocks in northwestern Utah.

In the early part of Phase III sediments were derived, in large part, from erosion of the interior of the continent and spread westward into southeastern Utah and adjacent states onto the Colorado Plateau from uplifts of the ancestral Colorado Mountains. The pattern reversed, however, about midway through Phase III, with sediments then being derived principally from erosion of newly created mountains along the orogenic belt in the west (fig. I.5C). The coarse conglomerate of Red Narrows along U.S. Highway 50 and 6 or of the northern Gunnison Plateau are representative of the bouldery beds swept eastward out of the mountains of the Sevier orogenic belt. Coal and the marginal coal swamps of Carbon and Emery Counties accumulated where the coarse erosional debris from the mountain uplift interfingered with fine marine shales at the coastline through central Utah. Late in Phase III time, one could have gone by boat almost without interruption from the Gulf of Mexico through to the Arctic Ocean along the relatively narrow, shallow, seaway that developed in the subsiding trough east of the Sevier Orogenic belt. The end of Phase III is marked by the last marine invasion of the sea into Utah and the Colorado Plateau.

Phase IV begins with development of broad lake basins which covered much of the Uinta Basin and areas to the south in the High Plateaus. Thick sequences of oil shale or limestone and shales accumulated in these lake basins. The oil shale of the Uinta and Piceance Basins was deposited during Phase IV development. These broad lake basins (fig. I.5D) were apparently outlined by major uplifts that developed during Laramide time, in the latest Cretaceous and earliest Tertiary upwarping. The San Rafael Swell, Circle Cliffs Upwarp and the Monument Valley Upwarp all apparently had their origin during this particular period of crustal instability when large vertical movements of blocks of the Colorado Plateau outlined the major domes and basins which were often separated by the distinctive monoclines of the plateau.

Uplift during Phase IV essentially ended the accumulation of thick sequences of sediment in the Colorado Plateau of western North America. A broad phase of erosion was then initiated, during which the extensive Canyonlands systems that produced much of the scenery of this section of the Plateau were developed. During the Laramide disturbance, rocks which had accumulated initially at sea level were uplifted to 10,000 to 15,000 feet above sea level, and into these uplifted blocks the canyons are now being carved. The Uinta Basin Section or the High Plateaus Section of the Colorado Plateau are remnants of these uplifted blocks. Sediments which initially covered much of the lower lying sections, as in the Navajo or Canyonlands sections, have now been stripped away and transported to the sea.

Phase V is principally the major volcanic phase (fig. I.5E) of the Colorado Plateau and lasted for approximately 25 million years. During the Oligocene major volcanic eruptions occurred around the margin of the plateau, for example, the volcanic rocks of the Marysvale region, of the High Plateaus.

Figure I.5. Six phases of the structural evolution of Utah and the Colorado Plateau (Modified from L.F. Hintze, 1973, Geologic History of Utah: Brigham Young University). *A*, Phase I, Late Cambrian to Devonian. Thick sequences of rocks accumulated in the miogeosynclinal area, while relatively thin sequences accumulated in the shelf or platform area. The present Colorado Plateau is in this area of thin layers. *B*, Phase II, Mississippian to early Triassic. Thick sections of rocks were deposited in the Paradox and Oquirrh Basins. Clastic debris was eroded from the bordering Uncompahgre, Emery, and Defiance uplifts. The uplifts and basins had a northwest-southeast trend. *C*, Phase III, Late Triassic to earliest Tertiary. The miogeosynclinal area northwest of the Colorado Plateau was folded and faulted during the Sevier Orogeny. The resulting mountains shed clastic debris into the seaways and subsiding area to the southeast. Coarse conglomeratic fans of the mountains piedmont graded eastward into coal swamps and marine deposits during the latter part of this phase. *D*, Phase IV, Latest Cretaceous to Eocene. Broad lakes covered much of the lowland between areas uparched during the Laramide disturbance. These uparched areas produced features like the Uinta Mountains, San Rafael Swell, and the Circle Cliffs, Monument, Kaibab, and Defiance uplifts. The Uncompahgre uplift was rejuvenated, in part. *E*, Phase V, Eocene to Oligocene. Major volcanic centers were produced during this phase of development. The Marysvale (M), Pine Valley (P), San Francisco (SF), and the San Juan Mountains (SJ) areas experienced early Tertiary volcanism. The LaSal (L), Henry (H), and Abajo (A) intrusions are shown by black circles and volcanic areas by a stippled pattern. *F*, Phase VI, Miocene to Recent. Uplift of the plateau and accompanying block faulting of the area to the west are the last major structural events that have affected the Colorado Plateau and the bordering provinces. Only the more significant boundary faults are shown on the map. The entire Basin and Range Province is complexly faulted.

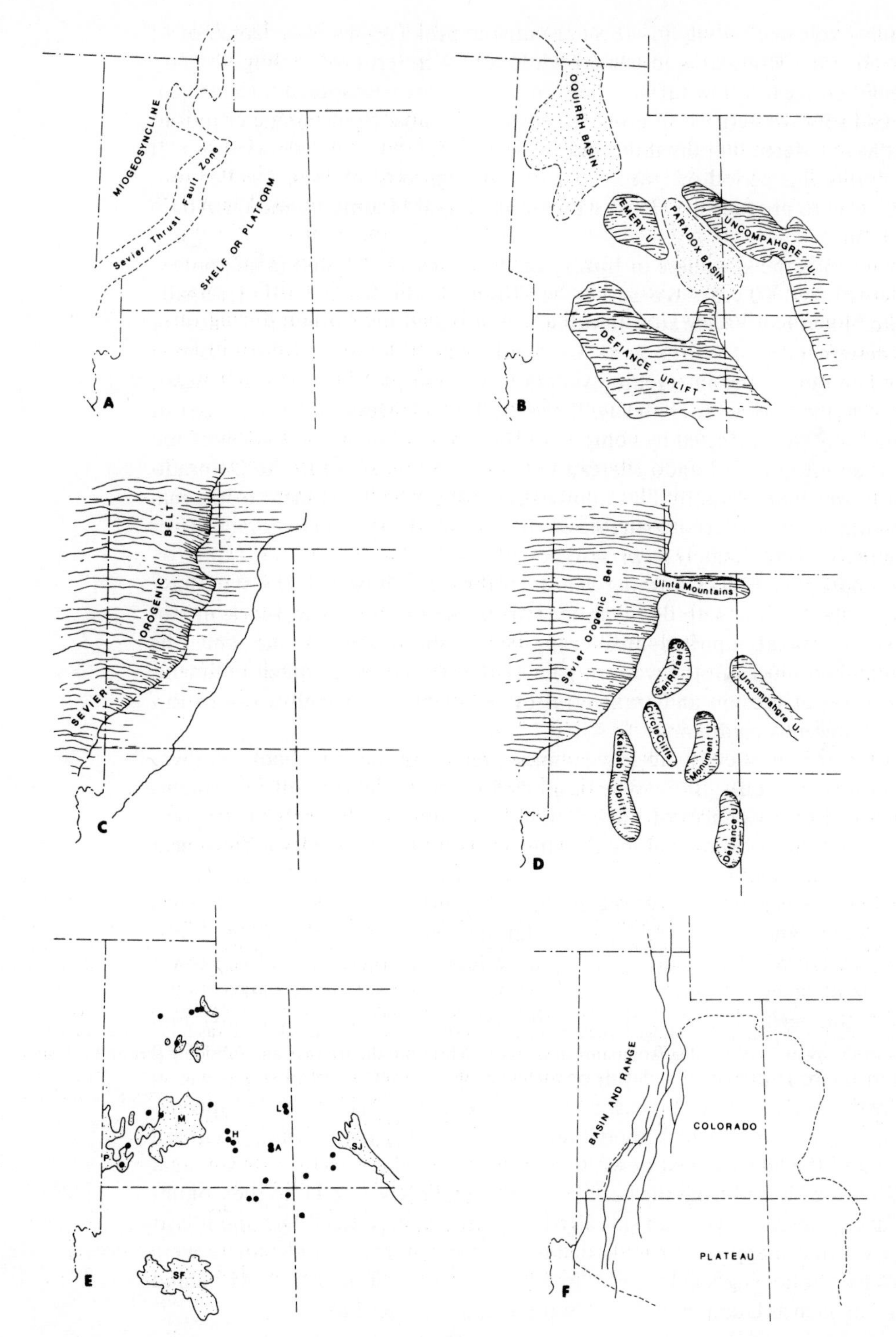
MIOGEOSYNCLINE
Sevier Thrust Fault Zone
SHELF OR PLATFORM
A
OQUIRRH BASIN
EMERY U.
PARADOX BASIN
UNCOMPAHGRE U.
DEFIANCE UPLIFT
B
SEVIER OROGENIC BELT
C
Sevier Orogenic Belt
Uinta Mountains
San Rafael
Uncompahgre U.
Circle Cliffs
Monument U.
Kaibab Uplift
Defiance U.
D
M
H
A
L
SJ
P
SF
E
BASIN AND RANGE
COLORADO
PLATEAU
F

Similarly volcanic activity in the San Francisco peaks region of Arizona, or of the San Juan Mountains in eastern Colorado were erupted during Tertiary time. Extensive ash flow tuffs, called ignimbrites, are widespread through central and southwestern Utah and into eastern Nevada. Some single eruptions must have covered literally thousands of square miles with hot glowing ash. It was during this period of igneous activity that igneous intrusions welled into the plateau to produce the Henry Mountains, LaSal Mountains and the Abajo Mountains.

Phase VI is the last phase of history of the plateau and lasted from approximately 25,000,000 years ago until now. This phase is characterized principally by prominent basin-and-range faulting and broad regional upwarp of most of the western part of North America. Faults have blocked out the isolated mountain ranges and intervening valleys in western part of Utah, in general, that region west of U.S. Highway 91 (fig. I.5F). The areas to the southeast of the highway were affected by comparatively minor faulting along some of the transition into the Colorado Plateau but the principal effect in the Colorado Plateau was that of uplift. This allowed even deeper entrenchment of the major streams, with the accompanying establishment of the main Colorado River drainage in approximately its present position.

Downdropped blocks forming valleys in the western part of the state have been partially filled with debris eroded from the uplifted fault block mountains. In general, uplifted areas experienced sharp erosion and only the downfaulted blocks in the western part of the state received much sediment. This phase of erosion and transferral of sediments from high areas to low areas is presently continuing.

One of the latest events to leave a major record in the guidebook area was the filling of the Lake Bonneville Basin during the glacial epics of the Quaternary (fig. I.6). Evidences of Lake Bonneville, such as its shorelines and flat-topped deltas, are exposed along the guidebook routes in the region from near Provo to the south in Utah Valley. To a lesser degree Lake Bonneville shorelines are evident in other regions farther south along U.S. Highway 91 in Juab Valley and in the vicinity of Fillmore. Notches produced by wave-generated erosion have outlined the upper levels of the lake and the lower broad fault valleys are floored by fine sediments swept out into the lake.

Lake Bonneville had a relatively cyclic history and at its highest maximum elevation the lake covered approximately 20,000 square miles in the western part of Utah. It obtained a depth of approximately 1,000 feet above the level of Great Salt Lake. At its highest point the lake rose to an elevation of approximately 5,200 feet, at which time it apparently overtopped a dam in southern Idaho and the lake drained down to approximately the 4,800-foot level, as a massive flood issued from the basin down the Snake River and into the Columbia River. The lake level then stabilized at approximately 4,800 feet and it is at this level that many of the wide deltas in Utah County and elsewhere were deposited. Following a climatic shift the lake then dessicated, leaving behind remnants such as Utah Lake, Great Salt Lake, and Sevier Lake.

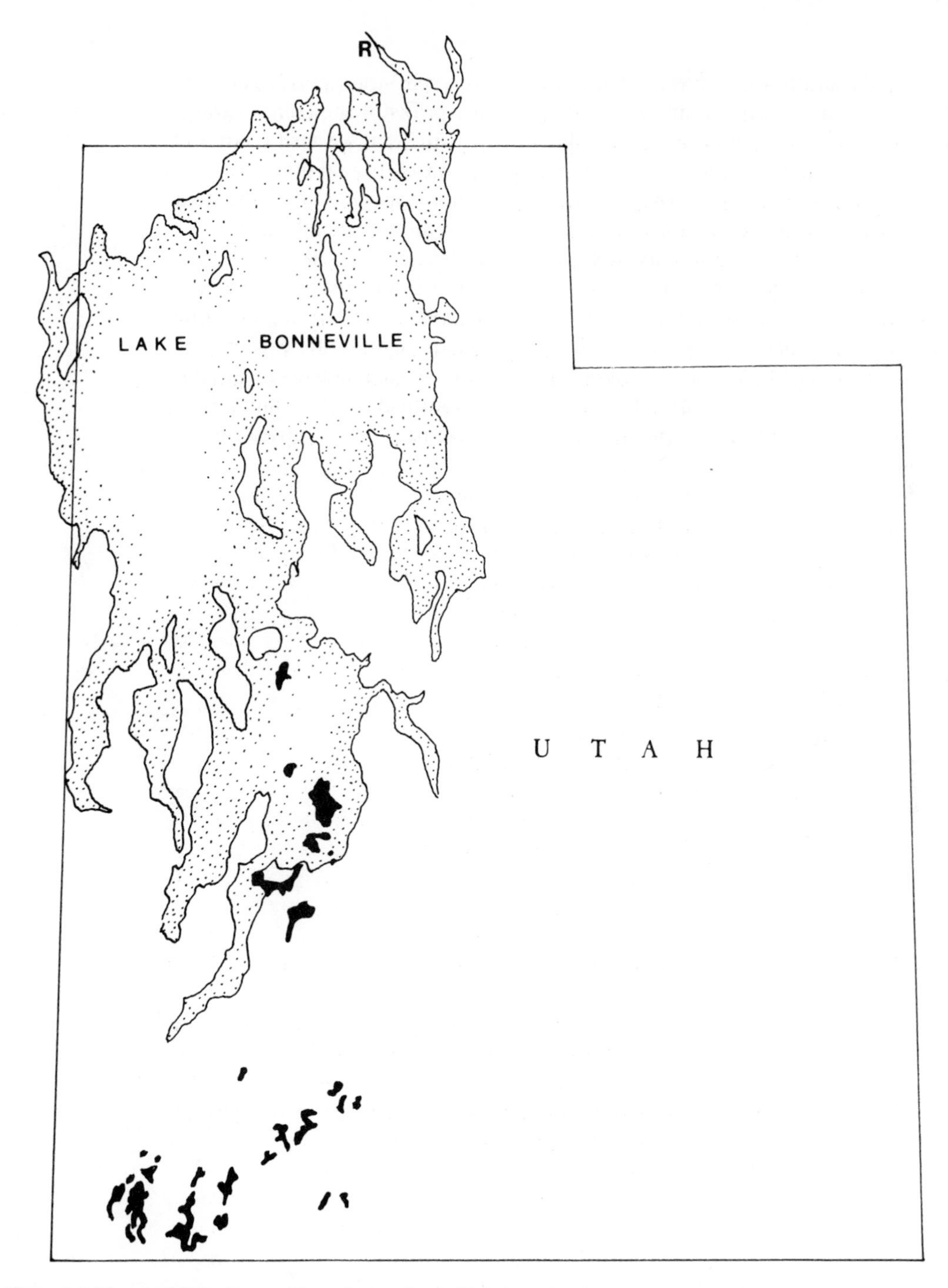

Figure I.6. Extent of Lake Bonneville, at its maximum high-water level (stippled area) and occurrences of Quaternary basaltic volcanic rocks in Utah (black). Part of the guidebook route is along the eastern shoreline of the lake and also through the volcanic fields in the central and southwestern part of the state. The lake drained northward through Red Rock Pass (R).

Some glaciation was also experienced by the Wasatch Mountains, Fish Lake Plateau, Wasatch Plateau, and some of the higher peaks of the Henry Mountains and LaSal Mountains at essentially the same time that western Utah was occupied by Lake Bonneville. Following dessication of the lake and the accompanying total retreat of the mountain glaciation topography of the area assumed essentially the relationships that we see now along the highways.

In a few areas volcanic activity was also intermittent during Lake Bonneville time. Some of the youngest volcanoes within the guidebook area are those near Fillmore where some are younger than Lake Bonneville. These young volcanic eruptions are principally basalt, in contrast to the more andesitic earlier Tertiary volcanic accumulations. Essentially the only more recent change was produced by occupancy of the area by man and the modification that mineral exploration, community development, and agriculture has superimposed on the landscape.

Segment 1

0.0 Junction of U.S. Highway 50-6 with Interstate Highway 15 Access Road to U.S. Highway 89 Northeast of Spanish Fork. The road is across Lake Bonneville prodelta and lake-bottom clay. A short distance to the east and southeast the gentle slope of the foreset beds of the deltas of Spanish Fork River and Hobble Creek limit the lake clays. A series of springs issue at the contact where porous sand and gravel overlie the impervious clay and mud.

0.8 Access road crosses junction with U.S. Highway 91 northeast of Spanish Fork. Southeast of the intersection the highway climbs up the front of the deltaic foreset beds.

1.2 Cross overpass east of north end of Spanish Fork. Here the foreset beds of the Provo-level delta into Lake Bonneville are expressed as the slope zone both to the southwest and northeast. The highway climbs onto topset beds of the delta southeastward.

3.4 Spanish Fork Peak on Maple Mountain is the high promontory to the east. Faceted spurs with triangular terminations show on the mountain (fig. 1.1) and suggest recurrent movement along the Wasatch Fault at the base of the mountain. Bonneville and Alpine Terraces of Lake Bonneville are inscribed into the base of the Maple Mountain. The level flatland at the highway was formed at the Provo level of the lake.

Figure 1.1. Triangular facets on Maple Mountain as seen from in the west near the mouth of Spanish Fork Canyon. Facets are the triangular terminations of the ridges above the Lake Bonneville shoreline, which shows in the center of the photograph as a horizontal break in slope. The highway is constructed on top of the Provo-level delta of Lake Bonneville.

Toward the south the light-colored outcrop at the mouth of the small canyon west of Spanish Fork Canyon is Diamond Creek Sandstone. The sandstone was lowered as part of a slice along the Wasatch Fault and gives some measure of the several thousand feet of vertical displacement on the fault.

4.2 Major bend on the road. Spanish Fork River is entrenched into the Provo-level terrace to the south. Gravel quarries are in deltaic sediments.

4.6 The highway climbs over a small-displacement fault scarp. The fault offsets the Provo delta surface and is marked by a prominent sag pond on the northern down-dropped block. The sag pond is particularly well preserved to the west, (fig. 1.2) but is obscured by the water tank and gravel quarries to the east.

Figure 1.2. Scarp of a post Provo-level minor fault, a branch of the Wasatch Fault, as seen west from Mile 4.6 in the first part of the section. The brushy escarpment is the fault scarp and the relatively barren pastureland in the immediate foreground is in the sag pond. This is the same stratigraphic surface as that at the top of the brushy escarpment and indicates the amount of displacement on the small fault. Paleozoic rocks in the background in the Wasatch Mountains are on the upthrown part of the main fault block along the Wasatch Fault, which here has a displacement in excess of 8,000 feet.

4.9 Cross main line of Denver and Rio Grand Railroad.

5.0 Junction of Access Road U.S. Highway 50-6 with U.S. Highway 89. Continue toward the southeast along the next part of Geologic Guide Segment 1.

0.0 Intersection of U.S. Highway 50 and 6 with U.S. Highway 89 near the mouth of Spanish Fork Canyon. The highway junction is on the Provo level of Lake Bonneville, an ancient Ice-Age lake that covered much of the desert area of western Utah. North of the junction a small fault has offset the upper surface of the Provo-level deltaic sediments. The flat area north of the fault, beyond one-half mile north of the junction, was deposited as the upper part of a delta of Spanish Fork River. When the delta was being built these flats were at the same level as the flat at the highway junction, but they have been dropped between 25 and 35 feet by subsequent faulting.

Higher levels of Lake Bonneville are marked by prominent lake terraces to the east and southwest of the highway junction. The most prominent terrace is at the Alpine level of the lake, at an elevation of approximately 5,200 feet. A less prominent higher terrace at the Bonneville level is inscribed into the flank of Maple Mountain at approximately 5,300 feet and marks the level of spillover of the lake into the Snake River and Columbia River. Following spillover, the rushing outpouring lake water quickly eroded the dam across the rim in southern Idaho and the lake level lowered to the Provo level at approximately 4,800 feet. The lake stayed at this level for some time until climatic changes reduced precipitation in the basin and the lake finally desiccated to its present remnants: Great Salt Lake, Utah Lake, and Sevier Lake. See the discussion of Lake Bonneville in the introductory part of the guide.

0.6 Overpass over spur railroad into Trojan Powder Company dynamite plant. Just beyond the overpass the front of the mountains marks the trace of the Wasatch Fault, one of the largest displacement and longest recently active faults of western United States. The fault has estimated displacement of 10 to 15 thousand feet near here, with the valley filled with 6 to 8 thousand feet of unconsolidated sediments. Only the youngest of these sediments are related to Lake Bonneville.

Bedrock of the mountains visible approximately 300 yards south of the overpass is of the Oquirrh Formation, one of the thickest formations in Utah (fig. 1.3). These rocks record the beginning of deepwater deposi-

PROVO AREA - SPANISH FORK CANYON

System	Series	Formation	Thickness (ft)	Notes / Fossils
PERMIAN	Wolfcampian	Oquirrh Formation (not shown to scale): Wolf-campian age Oquirrh 6200	19,000 (Hobble Creek Area) to 25,000 (Provo Canyon Area)	*Pseudoschwagerina*, *Schwagerina*
PENNSYLVANIAN	Virg	Pole Canyon Member 2000		*Waeringella*
	Mis	Lewiston Peak Member 2600		*Triticites*
	Desmoinsian	Cedar Fort Member 6200		*Wedekindellina*, *Fusilina*, *Chaetetes*
	A	Meadow Canyon Member 1100		*Profusilinella*
	M	Bridal Veil Falls Member 1245		*Endothyra*, *brachiopods abundant*
MISSISSIPPIAN	Chesterian	Manning Canyon Shale	1650	*Lepidodendron*, *Rayonoceras*
		Great Blue Limestone	2700	*Archimedes*, *Fenestella*
	Meramecian			*Apatognathus*
		Humbug Fm	930	*Fenestella*, *Polypora*
		Deseret Ls	850	*Gnathodus*
	Os	Gardison Ls	550	*Pseudopolygnathus*, *Syringopora*
	K			
D		Fitchville Fm	260	-white marker
CAMBRIAN	Middle	Maxfield Ls	600	*Kootenia, Spencia*, *Dolichometopus*
		Ophir Fm	250	*Glossopleura*
	L ?	Tintic Quartzite	1080	
PRE €		Mineral Fork Tillite	0-200	dolomitic clasts
		Big Cottonwood Fm	1000±	slate

System	Series	Formation	Thickness (ft)	Notes / Fossils
Q		Lake Bonneville Group: Provo Fm, Bonneville, Alpine Fm	0-200	
Ø		Moroni Volcanics	0-500	= Laguna Springs & Packard of Tintic
EOCENE		Uinta Fm	0-500	
		Green River Fm	0-5000	thickens eastward rapidly
		Flagstaff Fm	0-600	*fresh water snails*
CRETACEOUS		North Horn Fm	0-2500	coarse cong at Red Narrows
		Price River Fm	700	ANGULAR UNCONFORMITY
		Indianola Fm	2600	red mudstone, sandy shale, cg
JURASSIC		Morrison Fm	1900?	*dinosaur bone*
		Summerville Fm	275	brown mudstone
		Curtis Fm	235	greenish gray shale
		Entrada Ss	1270	tan silty ss
		Twin Creek Ls (Carmel)	1570	*Pinna*, *Pentacrinus*
		Nugget Ss	1450	cross-bedded, Navajo equivalent
TRIASSIC		Ankareh Shale	1530	red shale (Chinle equiv), brownish red sandy shale (Moenkopi)
		Thaynes Fm	1340	ls, red shale, *Meekoceras*
		Woodside Shale	315	
PERMIAN	Wordian	Franson Memb of Park City Fm	600±	UNCONFORMITY, *Punctospirifer pulchra*
		Meade Peak Sh M Phosphoria Fm	175	phosphatic shale
	Leon	Grandeur Member Park City Fm	880	*Dictyoclostus ivesi*
		Diamond Creek Sandstone	835	
	Wolfcampian	Kirkman Ls	1200-	*Pseudoschwagerina*

Figure 1.3. Stratigraphic section of rocks exposed in the Provo and Spanish Fork Canyon area in the southern Wasatch Mountains (from Hintze, 1973).

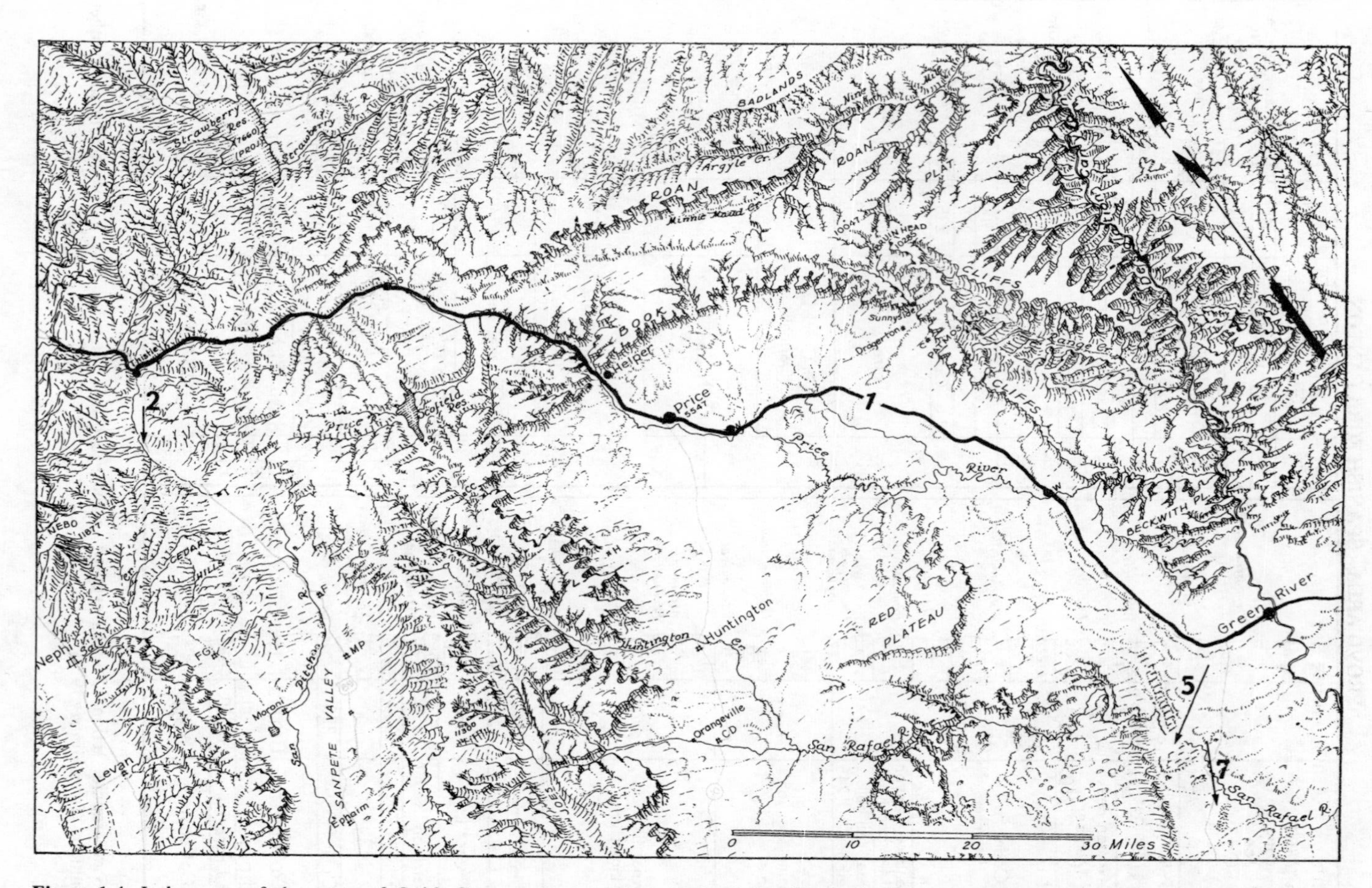

Figure 1.4. Index map of the route of Guide Segment 1, from the mouth of Spanish Fork Canyon southeastward to near Green River. Guide Segment 2 leads south from near Thistle and Segments 5 and 7 lead south and westward from near Green River. (Base map from Merrill K. Ridd)

tion in the subsiding Oquirrh Basin of Utah during Pennsylvanian-Permian time and the end of shallow-water fossiliferous limestone deposition.

U.S.Highway 89 and 50-6 have been constructed along the floor of Spanish Fork Canyon for several miles (fig. 1.4). The canyon is, in part, rejuvenated with the removal of Lake Bonneville-related sediments which partially filled the valley. Terraces along the margins of the canyon mark former levels of fill.

2.9 Strongly-jointed Oquirrh Formation limestone and sandstone is exposed in road cuts on the north. Silty and organic-rich layers which separate the thicker units have been burrowed and contain trails suggestive of a deepwater environment.

4.9 Quarry in tan Diamond Creek Sandstone. A warm sulphurous spring has been developed along the valley floor east of the quarry. The spring rises here at the crest of an anticline expressed in the sandstone. East of the spring area badly broken outcrops of the same sandstone are exposed in road cuts.

5.6 U.S. Highways 50-6 and 89 cross the Permian-Triassic boundary. Reddish easily eroded Triassic shale and siltstone to the north near the telephone line rest on ragged-weathering cherty Permian rocks to the west.

6.1 Junction of Diamond Fork road with U.S. Highway 89. The road cut on the north of the junction is in fossiliferous Triassic Thaynes Formation and the reddish slopes east of the junction are in the Ankareh Formation. Both of these formations are dipping to the east and are overlain along the skyline to the south by nearly flat-lying Tertiary rocks (fig. 1.5). The erosional surface between these two marks an angular unconformity and helps date the period of folding of the older beds.

The younger Tertiary rocks are soft and have flowed down into Spanish Fork Can-

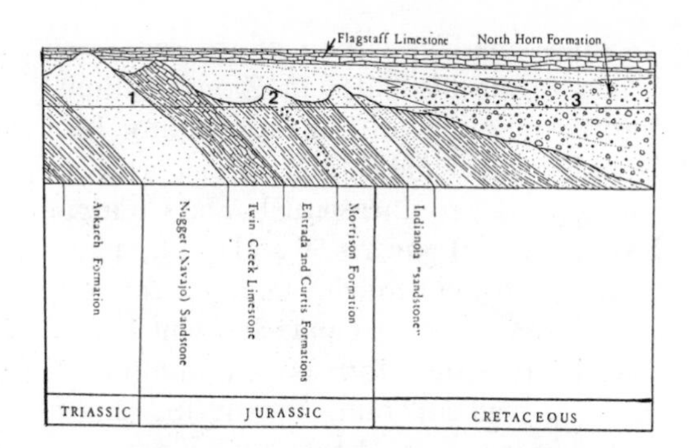

Figure 1.5. Generalized east-west structure section between Thistle (1) and Red Narrows (3). The Morrison Formation (2) is exposed at Mile 10.1. The horizontal line through the middle of the diagram is the approximate road level and shows the preunconformity rocks, below, dipping steeply toward the east with the Tertiary rocks, above, relatively flat lying.

yon on the south margin. This produces the hummocky rolling landscape visible above the railroad cuts in the sweeping bend of the canyon. A long tongue of landslide debris extends down through a gap in Navajo Sandstone on the west of the canyon 1.5 miles ahead, north of Thistle.

Diamond Creek is now nearly choked with debris swept down the canyon. The stream course has adjusted to a major increase in water volume produced by diversion from Strawberry Reservoir. As volume increases the formerly small meanders will increase to form large meanders that will be in equilibrium with the greatest stream flow. Such adjustment will result in major modification of canyon bottom meadows.

8.6 Thistle Junction: Separation of U.S. Highways 50 and 6, to the East, from U.S. Highway 89, to the South. High cliffs of Jurassic Navajo Sandstone form promontories north of the junction (loc. 1, fig. 1.5). Less prominent, in part fossiliferous, Twin Creek Limestone overlies Navajo Sandstone on the east and forms the small valley, ledges, and road cut cliffs to the east. Navajo Sandstone here is part of the same unit which forms the cliffs in Zions National Park, Arches National Park, Capitol Reef National Park, and in other scenic areas of southern and southeastern Utah. **Continue East on U.S. Highways 50 and 6.** For a continuation of the geologic guide south along U.S. Highway 89, see Guide Segment 2.

10.1 Jurassic Morrison Formation exposed in road cuts on the north (loc. 2, fig. 1.5). Fragments of bones and plant have been found here. Morrison Formation has produced most of the Jurassic dinosaurs known from western North America. The exhibit building at Dinosaur National Monument in northeastern Utah is centered on fossils in the Morrison Formation. High terraces to the south and north were probably related to a high level of Spanish Fork Creek during Lake Bonneville time.

10.8 Prominent very light gray Cretaceous sandstone ledges north of the highway are remnants of an upturned barrier island complex. Flat-lying Tertiary North Horn (red) and Flagstaff Formations (white) occur above tilted Cretaceous sandstone, above an erosional surface and angular unconformity (fig. 1.5). The North Horn Formation is composed of debris weathered from uplifted mountains to west which were created during the Sevier orogeny in Cretaceous time (see introductory section). This wedge of redbeds thickens to the east and is part of the thick conglomerate exposed in Red Narrows (fig. 1.6).

14.1 West entrance to Red Narrows (loc. 3, fig. 1.5). Conglomerate ledges here are composed of cobbles and pebbles of Precambrian and Paleozoic limestone and quartzite eroded from mountains of the Sevier orogenic belt in western and central Utah and were deposited here as part of alluvial fans or piedmont deposits. Lenses and filled stream channels are typical of the coarse textured units.

Figure 1.6. Conglomeratic exposures of the tongues of conglomerate of the North Horn Formation, like that shown in fig. 1.5., as seen up a small canyon toward the north at the west edge of Red Narrows, at approximately Mile 13.5.

15.8 Spring with Fountain at Roadside Rest Area, east end of Red Narrows. Red conglomerate fills a faulted square-profiled channel across the railroad tracks and stream to the southeast of the fountain. Tufa cones on the north (fig. 1.7) have developed at the main springs and are composed of calcium carbonate. Layers of tufa-coat leaves

Figure 1.7. Small tufa cones associated with carbonate-charged springs near the eastern end of Red Narrows at Mile 15.8. Red bark birch and other plants grow around the wet margin of the spring deposits.

and twigs of red-barked birch and other plants nearby. These springs are located along minor fractures associated with a major fault system that is present a short distance to the east. These faults allow water to well up from below or have allowed access to water percolating into the openings from adjacent porous beds.

16.5 Outcrops of Flagstaff Limestone in low ledges at the northern margin of the bend, across the old road. Pisolitic and oncolitic algal balls form onionlike masses in some of the limestone ledges. These algal balls probably accumulated along the shore or beach of Lake Flagstaff and enlarged by back-and-forth water motion and precipitation of calcium carbonate.

Carbonaceous shale and siltstone east of the Flagstaff ledges, near the low double road cuts on U.S. Highway 50-6, are marsh deposits associated with late stages of Lake Flagstaff or early stages of Lake Green River. Crocodiles, turtles, and many clams and snails occur with much broken or macerated plant debris in these beds.

17.2 Baer's Bluff. Lower beds of the Green River Formation are well exposed here. Greenish very low-rank oil shale, light gray limestone, tan papery laminated dolomite, and tan sandstone are part of deposits of Eocene Lake Green River. Fossil fish, snails, clams, ostracods, turtle fragments all occur in some of the limestone and silty beds, but not in the greenish rocks. These rocks show a transition from deltaic marshy environments in the lower part to open lake environments in the upper part. Algal coated logs form mounds in the middle of the cut. The canyon through which the highway is constructed for the next several miles is in the lower Green River Formation. Small faults offset some of the white limestone beds in dramatic fashion in railroad and highway cuts to the east (fig. 1.8).

23.8 Railroad Overpass. Excellent exposures of Green River Formation and its cycles show near the overpass. Two miles east of the overpass (Mile 25.7) small faults are well exposed in railroad cuts north of the highway and east of a service station- motel complex.

Figure 1.8. Small-displacement fault in the Green River Formation on the north side of road cuts at Mile 19.4. The prominent white limestone on the right shows fault drag into the fault. The same bed is exposed to the left of the fault as the two light outcrops in the lower left of the photograph. Darker beds are very low grade oil shale and green shale and siltstone that make up much of the lower part of the Green River Formation.

26.7 Roadside Rest Area and Junction to Skyline Drive, a scenic Forest Service summer dirt road along the crest of the Wasatch Plateau to the south. Excellent bird tracks have been collected in platy beds along the west flank of Starvation Creek, south of the rest area.

Erosion of sediments of the western edge of the Colton delta forms the valley through which the highway is constructed beyond Starvation Creek. The Colton Formation is a series of interbedded reddish shale and siltstone in this part of the delta. This part of the section, plus part of the underlying Flagstaff Formation, is susceptible to slumping when steeply tilted. Hummocky landslide and creep topography is particularly well shown on the south side of the canyon for the next four miles.

29.1 Railroad Overpass. Colton Formation is the reddish rocks near the valley bottom, with light greenish gray Green River Formation exposed along the canyon crest to the north. Hummocky topography to the south for a mile east and west of the overpass is produced by Colton slump masses. Some trees to the south have twisted trunks and have adjusted to downslope movement of their substrate.

31.1 Spanish Fork Canyon is narrowed here by landslide and slump mass from south. Step-type topography is related to dams formed by the toes of slumps, with flats formed by partially filled valley floors above the steep dams. The small stream has been displaced northward as the landslide mass moved out into the canyon.

33.8 Wasatch County—Utah County Line in Community of Soldier Summit. Strike valley is formed by softer Colton beds between more resistant Green River Formation on the north and Flagstaff Limestone on the south. Small dumps east of town are related to abandoned ozokerite mines and processing plants. Ozokerite is a purplish to black waxy solid hydrocarbon which was mined here and at other places between here and Colton, Utah to the southeast. The Soldier Summit Mine worked pockety deposits to a depth of 600 feet. Crushed rock was put into vats of water and ozokerite was skimmed off, then the skimmed fragments were melted to purify the wax.

36.7 Utah County—Wasatch County Line. Meandering White River occupies a subsequent valley in soft Colton deposits, some of which are exposed in gravel-capped terraces to the east. Indian Head (elevation 9,837 feet) is the high peak on the skyline to the east and is held up by Green River rocks.

39.7 Junction Utah State Highway 96 with U.S. Highway 50-6. Utah Highway 96 leads westward to Scofield Reservoir and Clear Creek. Scofield Reservoir is in a major graben, an elongate faulted trough. Clear Creek, Scofield, and Winterquarters were established as coal towns. Scofield was shipping coal as early as 1879. Winterquarters was the site of the worst mine disaster in Utah history when 206 men lost their lives in a coal-dust explosion there on May 1, 1900.

40.6 Minor junction to the south leads to old railroad siding of Colton. Approximately 1 mile beyond the side road the highway crosses a major fault (fig. 1.9) and then passes between double road cuts of the lower part of the Colton Formation which expose excellent channel-fill sandstone lenses cut into flat reddish siltstone and mudstone (fig. 1.10). Light-colored thin beds near the base of the cuts are fossiliferous lacustrine deposits over which the Colton delta was deposited.

Figure 1.9. Southeastward across the town site of Colton, a railroad switchyard in earlier days on the railroad. The prominent escarpment behind the buildings to the left is a fault-line feature associated with one of the major grabens in the crest of the Wasatch Plateau. Flagstaff Limestone caps the ridge to the left of the fault scarp and also forms the lower exposures at the extreme right of the photograph on the opposite side of the fault' on the downdropped block.

Figure 1.10. Cross section of the fillings of a meandering stream in the lower part of the Colton Formation in double road cuts at Mile 32.5. This is one of the three channel-filling sandstones exposed in the roadcut and has channeled into lacustrine and fluvial deposits.

41.5 Cross Price River and D&RGW Railroad Main Line Track. Flagstaff Limestone is exposed in gorge south of the highway and Colton beds are exposed to the north. The highway continues to the east through these same rocks.

44.6 Roadside Rest Area and Road Junction. The Utah State Highway heads east along the subsequent valley on the Colton Formation. U.S. Highway 50-6 heads southeastward down Price River Canyon. Flagstaff Limestone is exposed near the rest area, and forms the well-bedded cap of the stripped surface east of the canyon and rest area.

45.8 Utah County—Carbon County Line, at the lower end of deep double road cuts in North Horn Formation. North Horn rocks here are a mixture of stream and lake deposits and similar rocks are exposed for the next 3 miles down the canyon. Lower parts of the formation contain some thin beds of shiny black coal, as for example in cuts near Ford Creek at Mile 47.3.

48.1 Road junction to Price Canyon Recreational Area. Base of the North Horn Formation and top of the Price River Formation is placed at the red shale exposed in the road cut approximately one mile downstream from here. Below the red shale the highway continues in relatively massive sandstone and shale of the Price River Formtion. On some cuts, particularly at Mile 49.6, joints and interbedded sandstone and shale combine to produce an area of perennial slumping and road damage.

51.6 Scenic Turnout to View Castlegate (fig. 1.11). Castlegate Sandstone forms the 'gate'' ahead and also the vertical cliffs in the vicinity of the viewpoint. The sandstone is of deposits of sediment-charged braided streams draining across the piedmont, in front of the Sevier orogenic mountains, into the coastal area and shallow sea that was at that time at approximately the Utah- Colorado state line. The highway drops down section into the underlying Blackhawk Formation (fig. 1.12) after passing through the Castlegate.

53.4 Scenic Turnout and Geologic Stop at Overlook Near Tipple of North American

Figure 1.11. View toward the south of The Castlegate. Base of the Castlegate Sandstone is exposed in the shaly carbonaceous breaks that form the dark band in the road cut on the right. The very lenticular bedding in the Castlegate Sandstone is typical of the formation over much of its extent in the Northern Colorado Plateau.

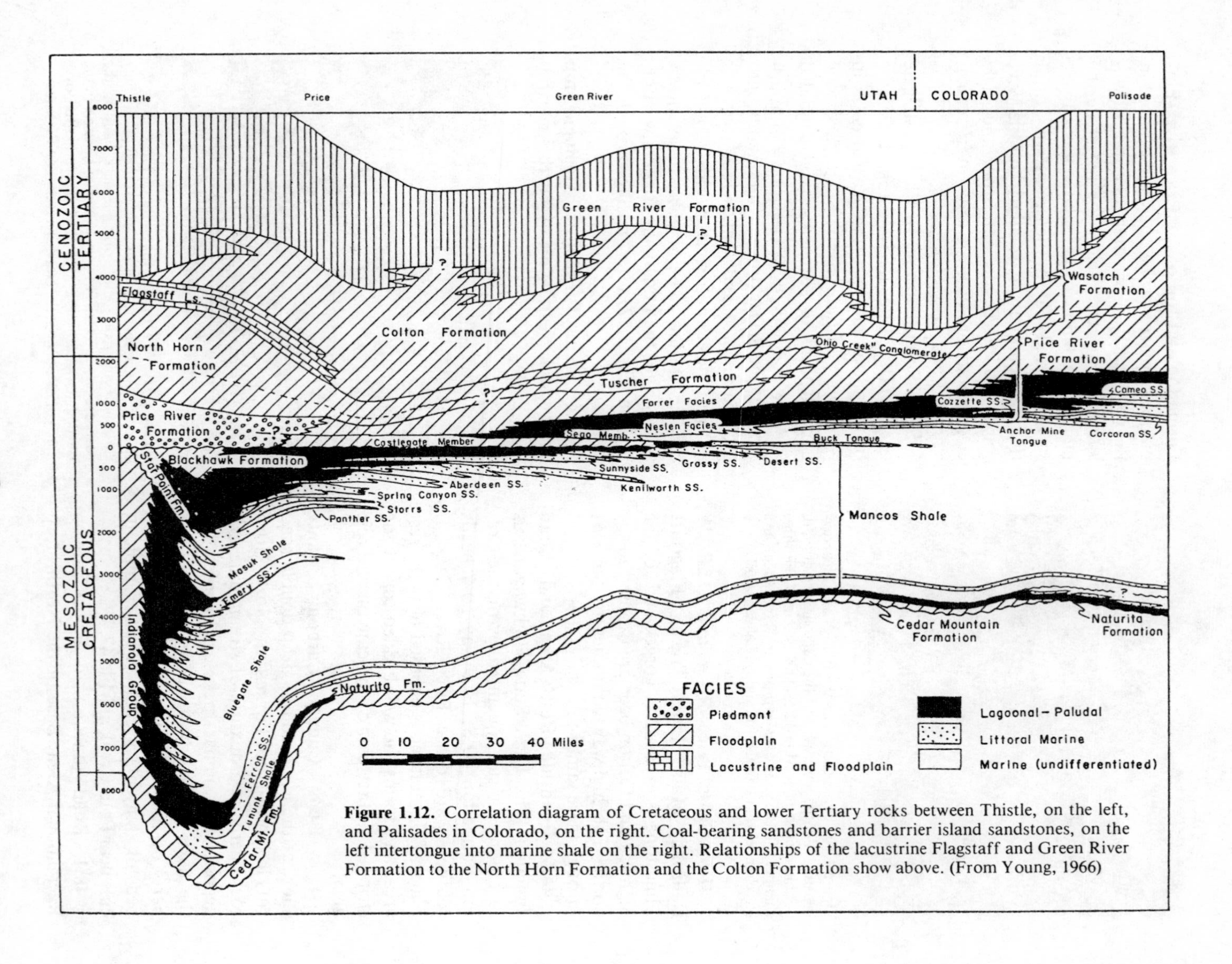

Figure 1.12. Correlation diagram of Cretaceous and lower Tertiary rocks between Thistle, on the left, and Palisades in Colorado, on the right. Coal-bearing sandstones and barrier island sandstones, on the left intertongue into marine shale on the right. Relationships of the lacustrine Flagstaff and Green River Formation to the North Horn Formation and the Colton Formation show above. (From Young, 1966)

Coal Corporation at Castlegate. Blackhawk Formation (fig. 1.12) is the major coal-producing rock unit in central and eastern Utah. It erodes to low ledges and slopes alternating with more massive light colored sandstone cliffs. The thicker sandstone units are barrier island accumulations and coal accumulated in coastal marshes and swamps behind these islands (fig. 1.13). The prominent white-capped sandstone behind the tipple, on the east side of the canyon is an example of one such barrier island deposit. The top of this sandstone is exposed in the road cut on the west of the highway (figs. 1.14, 1.15) and is overlain by a coal seam approximately 3 feet thick that accumulated in the marsh behind the island. The coal was buried by the marsh-filling stream deposits as the barrier coastline was crowded eastward by sediments being dumped in from the

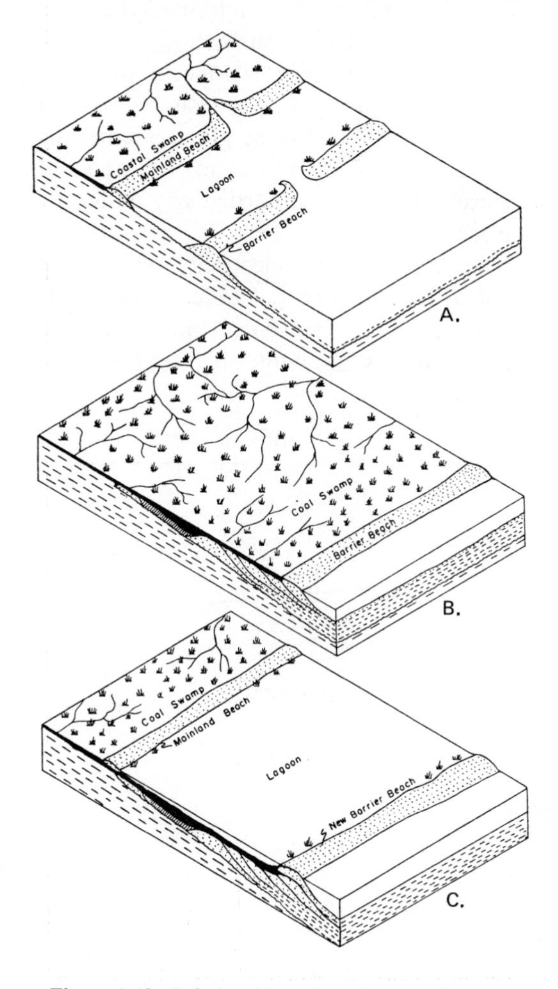

Figure A. Lagoon developed between two barrier beaches that are formed as a result of sediment being dumped into the Mancos sea and longshore drift of sediments along the coast from the northeast to the southwest.

Figure B. Development of coal swamps with filling of the lagoonal region behind the younger barrier beach sequence as part of the general regression of the coastline during the Cretaceous.

Figure C. With renewed subsidence a new lagoon may form on top of the now submerged coal deposits resulting in brackish water or shaly beds being deposited on top of the sandstones and development of a new high barrier beach sequence, at the beginning of a new coal accumulation cycle.

Figure 1.13. Relationships of coal accumulations and barrier beach developments along the Cretaceous shorelines as an example of the sedimentary model for Cretaceous coal accumulation in the Northern Colorado Plateau. (From Young, 1966)

Figure 1.14. High roadside exposures of the lower part of the Blackhawk Formation at the scenic turnout near the now-abandoned town of Castlegate at Mile 53.4. The lower sandstone beyond the cars is the top of the Aberdeen Sandstone and is overlain by the Aberdeen Coal, here approximately three feet thick. Additional lenticular lagoonal sandstone, coal, and argillaceous Carbonaceous accumulations that are typical of the late swamp filling occur higher in the roadcuts. The bleached upper part of the Aberdeen Sandstone is typical of some of the major barrier island sandstones that are overlain by coal.

mountains to the west. A lower and older barrier island and coal sequence is exposed in the deep double road cuts to the south (fig. 1.15).

Pinkish rocks to the northeast and near the top of the road cut on the west are the result of oxidation of iron-bearing minerals and development of clinker produced by burning of coals in place. Most burning appears to have been prehistoric and perhaps started by forest fires or lightning where the coal was exposed at the surface.

The Castlegate mine connects with the Kenilworth mine and as such is one of the largest coal mines in the state. Numerous dinosaur tracks are found in the lagoonal deposits of the mines, particularly on top of the coal beds.

The coal-burning electricity generating plant of Utah Power and Light Company is in the canyon to the south (fig. 1.16). It is one of the few mine-mouth fed generating plants in the West.

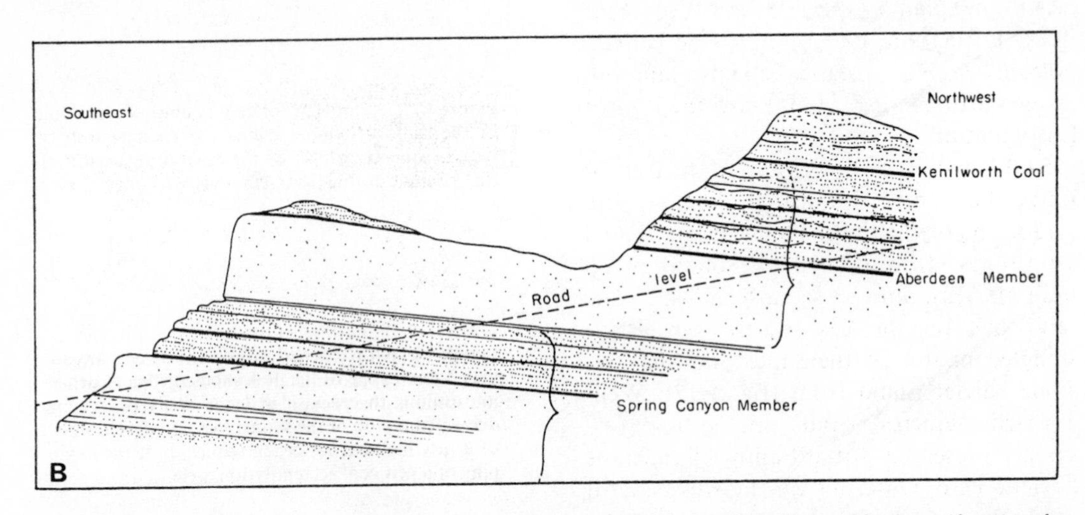

Figure 1.15. Generalized cross section of the exposures seen at a road cut at Mile 53.4 and in the road cuts to the south showing the characteristics of a basal barrier island sandstone, coal, and receeding lenticular, irregularly bedded sandstone and coal in the upper part of the member. The dashed line is the approximate road level of the highway. Vertical scale here is considerably exaggerated to show the stratigraphic relationship.

Figure 1.16. View southeastward over the lower part of Castlegate showing the "white-capped" Aberdeen Sandstone Member beyond the tipple and the power plant in the eastern canyon wall at Mile 53.4. The smoke plume is from the electricity generating plant of Utah Power and Light Company in the canyon to the south. The town of Castlegate formerly extended for approximately one mile up canyon from here but has been recently removed to allow coal development in the canyon bottom.

53.9 Junction of Utah State Highway 33 with U.S. Highway 50-6 west of steam-generating plant. Upper beds of the Star Point Formation are visible in the canyon bottom a short distance downstream opposite the mine waste dump, near the Port of Entry Station.

54.0 A short distance beyond the Port of Entry the Panther Sandstone Tongue (fig. 1.120, one of the lower barrier-island-based sandstones, forms distinctive outcrops on both the east and west canyon wall. The sandstone on the east portrays well the shingled nature of the sandstone and siltstone barrier island front (fig. 1.17). Well-exposed, numerous, and variable trace fossils are preserved in the Panther Sandstone Tongue here. Outcrops on the west side of the highway are most accessible for study. Panther Sandstone overlies the Mancos Shale, one of the thick gray Cretaceious shales of the Mountain West. U.S. Highway 50 and 6 is constructed in large part on Mancos Shale or alluvial valley fill over Mancos Shale from here to beyond Grand Junction in western Colorado.

56.5 Junction and Railroad Crossing in Northern Helper. Beheaded pediments which are cut across gently dipping Mancos Shale show beyond town to the east. They are armoured with gravel derived from the overlying sandstone units. Cretaceous units show in The Helper Face to the north (fig. 1.18).

59.7 Cross Price River. Garley Canyon Sandstone, a lower tongue of sandstone in the Mancos Shale, caps the bluff on both sides of the highway. Exposures of the sand-

Figure 1.17. Exposures of the Panther Sandstone at the type locality on the eastern side of Price Canyon as seen from approximately Mile 46.0. The Panther Sandstone is the prominent ledge and recess cliff-forming unit in the foreground. Within the exposure the gently inclined beds are typical of the shingled appearance of the unit which was deposited as a barrier island migrated from north to south along the coastline in the Cretaceous. Younger Blackhawk Formation and Star Point Formation form the white-capped sandstone cliffs and slope zone in the head of Panther Canyon.

Figure 1.18. View northward from west of Helper, at approximately Mile 48.0, toward The Helper Face. Lower slopes are on the Mancos Shale and the first prominent ledge is the silty remnant of the Panther Sandstone. Massive cliff-forming units above, near the top of the sheer part of the face, are in Blackhawk Formation and show rather characteristic coal and barrier island facies of the formation.

stone here are of the outer seaward part of a barrier island sandstone like the massive beds seen near Castlegate and like the shingled Panther Sandstone in the mouth of Price Canyon. The Garley Canyon Sandstone thins out into silty shale in a short distance to the east. Gravel-capped pediments are developed along Price River above poorly exposed gray Mancos Shale. The Book Cliffs rise above Price River Valley to the west and northeast, and mark the boundary between the northern and western high plateaus provinces and the canyonlands section of the Colorado Plateau.

64.5 Junction of Utah State Highway 10 with U.S. Highway 50-6 at the west edge of Price business district. Price City Museum with natural history exhibits is in the Carbon County Courthouse on the north side of the highway 0.2 miles east of the intersection.

East of Price the highway continues on terraces of the Price River that are cut into Mancos Shale. Toward the east an extension of the Book Cliffs forms the southern and southwestern limit of the Tavaputs Plateau which is capped by Green River Formation and younger rocks. Red Plateau and Cedar Mountain, at the northern end of the San Rafael Swell, are on the skyline to the south. Toward the southwest the southern extension of the Book Cliffs defines the boundary between the high Wasatch Plateau and the lower Canyonlands section of the Colorado Plateau.

71.5 Wellington Community Park. Naturally occuring CO_2 recovered from the Navajo Sandstone in Farnham Dome, 5 miles east of town, is converted here into "dry ice." The plant south of the highway at the east end of town supplies much of the "dry ice" demand in the state. A coal washing plant built along the Denver & Rio Grande right-of-way southeast of town is on the floodplain of Price River. East of town the highway continues across lower Mancos Shale and climbs onto the northwest flank of Farnham Dome.

76.0 Beginning of Cat Canyon on the northwestern margin of Farnham Dome. Thick Ferron Sandstone forms the resistant unit which caps the walls and forms the prominent cuesta around the northern and western part of the dome. Two small reverse faults cut the Ferron Sandstone and are particularly well exposed on the north side of the road (fig. 1-19). Gray Mancos Shale is exposed below the Ferron cap in the eastern part of the gap. The shale forms a subsequent valley around the dome between the resistant Ferron and Dakota Sandstones.

Figure 1.19. View eastward along U.S. Highway 50 and 6 from approximately Mile 67.0 showing fault-repeated exposures of the Ferron Sandstone in Cat Canyon. These steeply dipping rocks are on the west side of Farnham Dome and show three nearly complete sections of the Ferron Sandstone on fault-isolated blocks which have dropped down to the east. The massive sandstone which caps the exposures in the gap, near the bend in the road, is the same as that which forms the prominent exposures beneath the juniper cover at the left of the photograph.

77.2 Primitive side road to the south is along the crest of the dome and leads to the CO_2-producing well near the knob of reddish Cedar Mountain Formation exposed in the core of the fold to the south. Dakota Sandstone is exposed in highway cuts east of the side road junction and beyond the fence south of the road.

Ferron Sandstone nearly pinches out over the crest of the structure so that it forms only a silty, low, rounded discontinuous cuesta on the east flank where the highway crosses its trace.

80.7 Road Junction of Utah State Highway 123 with U.S. Highway 50-6 East of Grassy Trail Creek Crossing. Utah State Highway 123 leads east to Sunnyside, a coal mining and former coke town. Coal is mined from a Cretaceous barrier island-lagoon sequence which is similar to that exposed near Castlegate, although it is slightly younger here.

The largest known deposit of asphaltic sandstone in the United States occurs north of Sunnyside on the high part of the West Tavaputs Plateau. It is in beds 10 to 350 feet

Figure 1.20. View eastward through juniper woods to the western edge of the West Tavaputs Plateau, across coarse mud flow debris which is blanketing a pediment surface cut across Mancos Shale. Blackhawk and Castlegate Formations form the low ledges and the prominent ridge in the middle distance. Green River beds cap the plateau on the skyline.

thick in upper Colton and lower Green River beds and has been mined since 1892, although major development didn't take place until 1928.

U.S. Highway 50-6 continues southeastward over pediments and pediment gravel and through cuts in gray Mancos Shale. Some of the pediment-armouring gravel is composed of coarse blocks and boudlers brought down from the escarpment and canyons to the east by mudflows (fig. 1.20). Several levels of pediments are developed, each adjusted to a former position of Price River and its tributaries.

86.7 Carbon County-Emery County Line. The road continues in Mancos Shale or over pediment surfaces on the shale. Upper Mancos Shale is exposed in badlands along the cliffs to the east below coal-bearing Blackhawk Formation (fig. 1.21), massive Castlegate Sandstone, and more ledge-and

Figure 1.21. View eastward from the highway to badland topography carved on upper beds of Mancos Shale. Cliffs above are in Blackhawk Formation and Castlegate Sandstones and are part of the nonmarine nearshore and coal swamp accumulation associated with withdrawal of the Mancos Sea.

slope-forming Price River Formation that caps the first series of cliffs. Younger rocks up to Green River Formation are exposed in the second series of cliffs and ledges up to Patmos Head (elevation 9,841 feet) the high promontory on the rim of the West Tavaputs Plateau.

Cedar Mountain, to the west, is at the northern end of the Triassic and Jurassic rim of the San Rafael Swell, a large domal structure that dominates much of the geology of east central Utah.

103.5 Post office and service station at **Woodside.** A CO_2-driven geyser (fig. 1.22) erupts from a water well drilled by the railroad in 1910. The water well spouted as much as 75 feet into the air initially but escape of gas has lowered the driving pressure and height of the geyser eruption.

Figure 1.22. Woodside geyser at Mile 94.5 erupts approximately every hour from a well drilled by the railroad in the early 1900s. The photograph shows the last phase of eruption of the CO_2-driven geyser.

East of Woodside the Castlegate Sandstone forms the uppermost massive cliff. A lower massive sandstone in the Blackhawk Formation exposed along the gorge of Price River is subdivided into three whitecapped sandstone beds as the unit is traced northward. Each of these is capped by a coal bed in the lagoonal part of the section. Mancos Shale is exposed near the base of the Book Cliffs escarpment and the highway continues southward over the lower part of the shale beyond the crossing of Price River.

111.4 Overpass Over Main Line of Denver & Rio Grande Railroad. San Rafael Swell to the west exposed lower Cretaceous and Jurassic beds in cuestas along the east flank. Ferron Sandstone forms the low cuesta immediately west of the flat and the road.

117.6 Bend in the highway. The Reef of the San Rafael Swell to the south and southwest is held up by Triassic Wingate, Kayenta, and Navajo Sandstones where the beds dip steeply off the eastern flank of the uplift. U.S. Highway 50-6 continues through gray Mancos Shale. Beckwith Plateau, to the east, is capped by Castlegate Sandstone which here begins to develop thin coal beds and a lagoonal facies, equivalent to a barrier island sandstone to the east and to piedmont braided stream deposits to the west.

121.2 Route Separation. The road ahead leads eastward toward Green River on eastbound Interstate Highway 70, and that to the west (right) leads to Interstate Highway 70 westbound toward Salina. Both routes continue in the lower Mancos Shale. See Guide Segment 5 for a continuation westward along Interstate 70. Segment 1 continues ahead on eastbound Interstate 70.

121.5 Bridge over Interstate 70 on interchange. The highway now swings to the east in Mancos Shale, parallel to the southern margin of Beckwith Plateau. Castlegate Sandstone caps the plateau above ledges and slopes of upper Blackhawk Formation and the silvery gray slopes of Mancos Shale.

125.1 Downtown **Green River, Main Intersection.** The gorge of Green River north of town separates the East and West Tavaputs Plateaus. Powell gave the name Desolation Canyon to this stretch of the river gorge across the southern part of the Uinta Basin where the river cuts through the Roan Cliffs and Book Cliffs.

125.7 Emery County—Grand County line marking a former position of the midchannel for the Green River before it abandoned this meander.

126.4 Center of the Bridge Over the Green River.

127.6 Double road cut in silty beds of the Ferron Tongue in the lower Mancos Shale. To the east at the crossing of Brown's Wash the fossiliferous dark gray laminated part of the member is exposed to the south. To the north the lower cliff-forming part of the Book Cliffs escarpment is made up of Mancos Shale up to Castlegate Sandstone. These rocks are overlain by the valley- forming Buck Tongue of the Mancos Shale and by deltaic beds equivalent to the Price River and North Horn Formations in Price Canyon.

129.1 Underpass Beneath D&RGW Railroad. A short distance east of the underpass a side road to the south leads to Crystal Geyser, a CO_2-driven well and geyser. It puts on a spectacular display approximately every 4 hours.

132.7 Cross a small gully. The highway continues in Mancos Shale with a characteristic cap of rusty brown gravel derived from the plateau to the north. The white coating here and there is caliche of calcium sulphate brought to the surface by ground water and left behind by evaporation.

136.3 The road rises to an upper pediment surface. To the north the prominent high cliff in the immediate vicinity is capped by the Castlegate Sandstone. Lower lenses of sandstone appear beneath the coal-bearing, but now very silty Blackhawk Formation. These lower sandstone beds show the structure within the Mancos Shale. The LaSal Mountains are visible ahead at about two o'clock and the Salt Valley Anticline is developed in the Castlegate Sandstone and Mancos Shale to the north. The telephone relay station ahead is on the Castlegate Sandstone.

145.0 Junction of U.S. Highway 160 with Interstate Highway 70 and U.S. Highway 50-6. For a continuation along U.S. Highway 160 southeastward toward Moab and Canyonlands and Arches National Parks see Geologic Guide Segment 6. Both highways continue in Mancos Shale, either to the east at the foot of the Castlegate-capped Book Cliffs, or southward around the nose and flank of Salt Wash Anticline.

Segment 2

0.0 Thistle Junction. Cross Bridge and Head South on U.S. Highway 89 (fig. 2.1). Gravel fill is exposed in road cuts to the east. Cross-bedded Navajo Sandstone is exposed on the west and north (fig. 2.2) and basal beds of Twin Creek Formation are exposed just short of the crossing over Thistle Creek and to the east.

0.4 Cross Thistle Creek to the west side of the creek and continue south in the valley carved in the lower part of the Carmel or Twin Creek Formation. High country across the canyon to the west is a carved slope of Navajo Sandstone. High to the east Flagstaff and North Horn Formations are exposed. Asphaltic sand is found in the Flagstaff Formation.

0.8 Quarry in Twin Creek Limestone to the west and bright red sandstones and siltstones of the upper part of the formation exposed in bluffs to the east. High-level stream terraces can be seen to the east in the same general vicinity, approximately 200 feet above the river. This terrace marks a point of adjustment of the streams of Thistle Creek and Spanish Fork Canyon and may be related to renewed faulting along the Wasatch Fault at the mouth of the canyon.

2.0 White Upper Cretaceous sandstone exposed to east in valleys. Flats are site of former town and mine of Asphaltum built to exploit asphaltic sandstone in Green River Formation.

2.6 Outcrops of coarse volcanic conglomerate in basal beds of the Moroni Formation (fig. 2.3). These are dark grayish green outcrops and contain a variety of altered volcanic rocks. These conglomerates and associated volcanic rocks rest with angularity on the older tilted rocks below. The same formation is exposed for several road cuts on the east side of the road, but the eastward dipping Mesozoic rocks are well exposed to the west. Thistle Creek here is in a subsequent valley along the lower edge of the volcanic rocks and is, in some respects, a resurrected valley.

3.5 Hoodoos are well developed in the Moroni Formation immediately to the west of the railroad bridge over Thistle Creek, on the southwest side of the canyon (fig. 2.4). The immediate canyon is carved into these volcanic rocks for some distance to the south.

4.3 Cuts through tuffaceous units in the Moroni Formation. Small springs issue to the south of the road from the base of the tuffaceous unit. This road cut is usually damp all year long.

4.8 Pull Off Road to West at the side road which leads to the ranches on the west slope of the canyon. The irregularly carved face of the quarry from which the algal-ball limestones of the North Horn Formation (fig. 2.3) were obtained can be seen to the southeast, at ten o'clock high on the skyline.

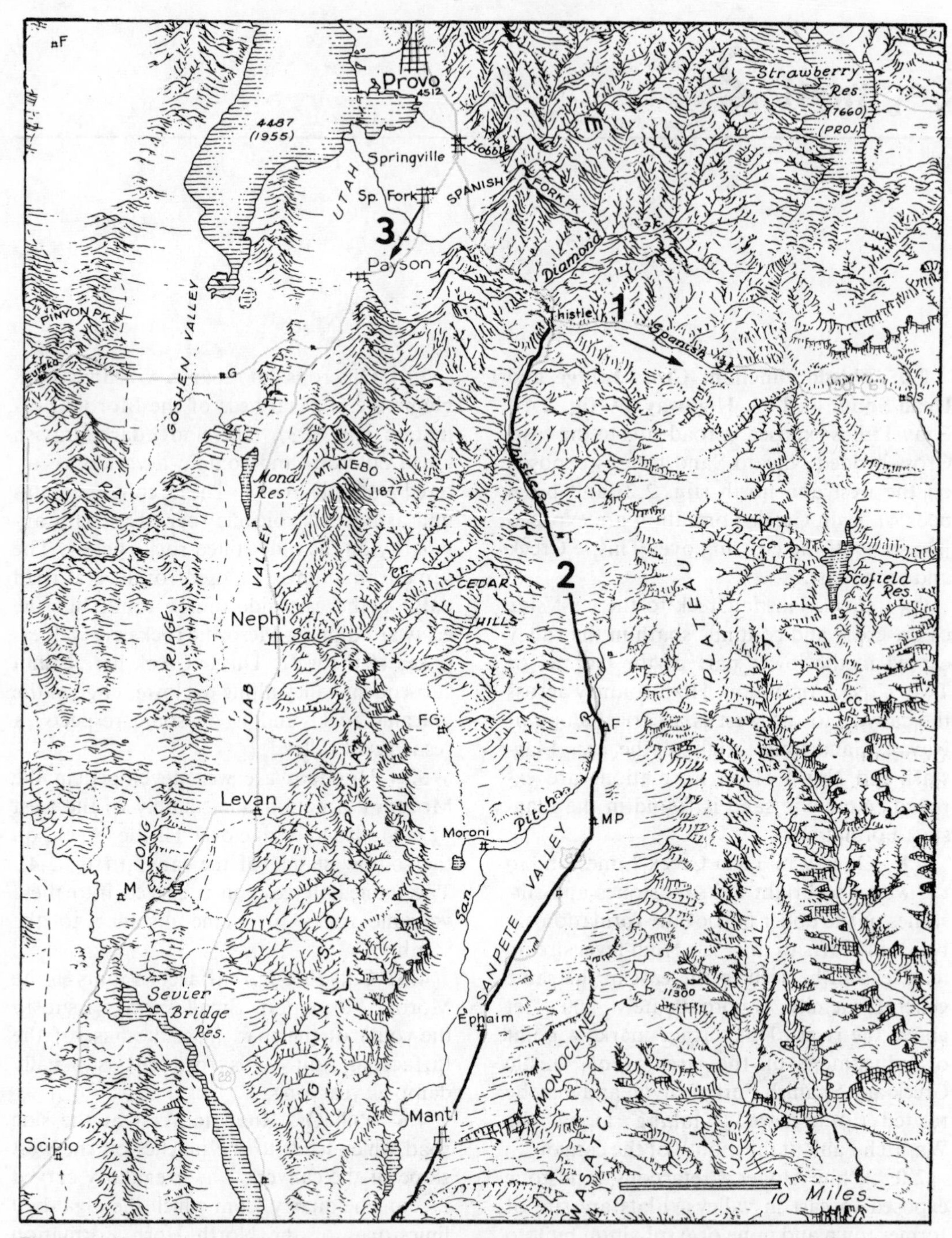

Figure 2.1. Index map showing the route of the northern part of Segment 2 from Thistle southward into Sanpete Valley. Route Segment 1 leads up Spanish Fork Canyon toward the southeast and Route Segment 3 leads southward along U.S. Highway 91 and Interstate Highway 15 from Utah Valley through Juab Valley.

Figure 2.2. Exposures of massive, cross-bedded, white Navajo Sandstone below slope-forming, well-bedded, argillaceous limestone of the Twin Creek Formation at Thistle Junction, the beginning of the route, at Mile 0.0.

This is the "Birdseye" marble quarry and was operated in the early 1900s. Some blocks of the "Birdseye" marble can be seen at the old loading site along the railroad track about a hundred yards to the south of the road junction.

6.3 Junction of the Road to the Small Community of Birdseye and Canyon of Bennie Creek. Mount Loafer (fig. 2.5) is the high point to the northwest, at about four o'clock, and is composed primarily of vertically dipping Oquirrh Formation. The high, flat, upland surface to the southwest, at two and three o'clock, is held up by Tertiary rocks which angularly overlie the folded Paleozoic formations. Excellent exposures of the pinkish tuffaceous volcanic rocks of the sequence can be seen east of the junction, near the Birdseye church house (fig. 2.6). The Birdseye marble quarry in the North Horn Limestone shows on the skyline to the east at nine o'clock. The road continues to the south at approximately the elevation of the intermediate terrace visible across the valley to the west.

7.3 Exposures of terrace gravels probably mudflow debris and alluvial fan outwash from the high country at the headwaters of the canyon to the east. The very coarse, irregularly sorted large blocks were probably rafted to here by mudflows. The coarse texture and angular shape distinguish these alluvial fan deposits from the rounded stream terrace material below.

9.2 The back slope of Mt. Nebo is visible to the southwest at the creek junction. Coarse terrace gravels and the underlying tuffaceous Moroni Formation visible to the east of the road.

10.0 Ranch house (east) and a small reservoir west of the road. Somewhat behind us to the northeast, red rocks of the Indianola and Price River Formations form the skyline. This is the type section of the Indianola Formation. Gray outcrops of the Moroni Formation, with terraces etched into their surfaces, can be seen across the valley to the west and southwest. These terraces were obviously adjusted to a higher position of Thistle Creek.

10.7 Tent Mountain and Brigham's Chair or the Horseshoe are visible directly ahead on the skyline. They are two distinctive peaks carved in the flat-lying Flagstaff rocks of the Wasatch Plateau. Roadside cuts are through terrace gravels which have partially filled Thistle Creek Valley and suggest two or three cycles of development for the valley.

13.9 Road drops to the floodplain of Thistle Creek. Gravel terraces show in road cuts to the east. Here Thistle Creek has entrenched only slightly into the broad fluvial floodplain of the valley. The floodplain is in part a depositional feature, having filled in a much more deeply excavated valley to the present relatively flat surface. Some abandoned meander channels of Thistle Creek can be recognized by more marshy areas throughout the generally flat valley floor. The road continues southward through terrace gravels.

PROVO AREA - SPANISH FORK CANYON

System	Series	Formation / Member	Thickness	Fossils / Notes
PERMIAN	Wolfcampian	Wolfcampian age Oquirrh 6200 (Oquirrh Formation, not shown to scale)	19,000 (Hobble Creek Area) to 25,000 (Provo Canyon Area)	*Pseudoschwagerina*, *Schwagerina*
PENNSYLVANIAN	Virg	Pole Canyon Member 2000		*Waeringella*
	Mis	Lewiston Peak Member 2600		*Triticites*
	Desmoinsian	Cedar Fort Member 6200		*Wedekindellina*, *Fusilina*, *Chaetetes*
	A	Meadow Canyon Member 1100		*Profusilinella*
	M	Bridal Veil Falls Member 1245		*Endothyra* *brachiopods abundant*
MISSISSIPPIAN	Chesterian	Manning Canyon Shale	1650	*Lepidodendron*, *Rayonoceras*
	Chesterian / Meramecian	Great Blue Limestone	2700	*Archimedes*, *Fenestella*, *Apatognathus*
	Meramecian	Humbug Fm	930	*Fenestella*, *Polypora*
		Deseret Ls	850	*Gnathodus*
	Os	Gardison Ls	550	*Pseudopolygnathus*, *Syringopora*
	K / D	Fitchville Fm	260	-white marker
CAMBRIAN	Middle	Maxfield Ls	600	*Kootenia, Spencia*, *Dolichometopus*
		Ophir Fm	250	*Glossopleura*
	L ?	Tintic Quartzite	1080	
PRE-€		Mineral Fork Tillite	0-200	dolomitic clasts
		Big Cottonwood Fm	1000±	slate

System	Series	Formation / Member	Thickness	Fossils / Notes
Q		Lake Bonneville Group: Provo Fm, Bonneville, Alpine Fm	0-200	
Ø		Moroni Volcanics	0-500	= Laguna Springs & Packard of Tintic
EOCENE		Uinta Fm	0-500	
		Green River Fm	0-5000	thickens eastward rapidly
		Flagstaff Fm	0-600	*fresh water snails*
CRETACEOUS		North Horn Fm	0-2500	coarse cong at Red Narrows
		Price River Fm	700	ANGULAR UNCONFORMITY
		Indianola Fm	2600	red mudstone, sandy shale, cg
JURASSIC		Morrison Fm	1900?	*dinosaur bone*
		Summerville Fm	275	brown mudstone
		Curtis Fm	235	greenish gray shale
		Entrada Ss	1270	tan silty ss
		Twin Creek Ls (Carmel)	1570	*Pinna*, *Pentacrinus*
		Nugget Ss	1450	cross-bedded, Navajo equivalent
TRIASSIC		Ankareh Shale	1530	red shale (Chinle equiv), brownish red sandy shale (Moenkopi)
		Thaynes Fm	1340	ls, red shale, *Meekoceras*
		Woodside Shale	315	UNCONFORMITY
PERMIAN	Wordian	Franson Memb of Park City Fm	600±	*Punctospirifer pulchra*
		Meade Peak Sh M Phosphoria Fm	175	phosphatic shale
	Leon	Grandeur Member Park City Fm	880	*Dictyoclostus ivesi*
	Wolfcampian	Diamond Creek Sandstone	835	
		Kirkman Ls	1200-	*Pseudoschwagerina*

Figure 2.3. Stratigraphic section of rocks exposed in the southern Wasatch Mountains in the Spanish Fork Canyon area (from Hintze, 1973).

Figure 2.4 Hoodoos in the Moroni Formation, Tertiary volcanic conglomerate, exposed on the west side of U.S. Highway 89 at approximately Mile 2.8. The coarse Moroni Formation rests with angularity across the steeply dipping Navajo Sandstone that forms the ledges in the background.

Figure 2.6. Well-bedded exposures of the Moroni Formation east of the church house at Birdseye, at Mile 6.3. The quarry on the peak on the skyline is in the North Horn Formation where oncolitic algal-ball limestone was quarried for finish stone near the turn of the century.

Figure 2.5. View westward from approximately Mile 4.8 across bouldery terraces of Thistle Creek to Loafer Mountain on the skyline. Loafer Mountain is carved in steeply dipping upper Paleozoic rocks and is flanked, particularly on the south or the left, by unconformably overlying Cretaceous and Tertiary rocks.

20.3 Historical Marker of an Indian Massacre that took place in the vicinity of the clumps of willows to the west. The gravel quarry is in tuffaceous gravels probably derived from the Moroni Formation.

20.5 Cross Utah County—Sanpete County Line and Enter into Thistle Valley. The high peaks to the east are on the northern part of the Wasatch Plateau. Thistle Valley separates the Wasatch Range, to the west, from the Wasatch Plateau to the east.

21.1 Junction of Side Road to Indianola, the small farming community to the east. Light gray and tan cones near the valley floor to the southwest at two to three o'clock are composed of tufa deposits related to warm springs in the area. Mormon baptisms were held in these small warm springs in the early days. High peaks of the central part of the Wasatch Plateau can be seen directly ahead on the skyline. The high part of the Wasatch Plateau over most of its surface to the east is held up by the Flagstaff Limestone. This is the same limestone that forms the high peaks of Musinia, Tent, and Brigham's Chair to the south. Flagstaff Limestone is exposed to the east along the crest of the Wasatch Monocline, in the Wasatch Plateau, but the Green River Formation occurs along the east edge of the valley in here. Green River Formation also forms the gentle cuesta ahead at eleven o'clock.

25.3 Excellent views of the Green River Formation to the southeast. One of the thicker limestone beds forms a prominent ledge along the margin of the cuesta. The cuesta also is held up by a resistant unit in the Green River Formation. Green River sediments accumulated in an ancient lake, probably near sea level, and have been subsequently uplifted to their present position.

27.8 Railroad Overpass Over D&RGW Railroad. Pull Off Road to Right Just Beyond Bridge. Road cuts are in rather characteristic exposures of the upper part of the Green River Formation. Abundant fish fragments, turtle fragments, small ostracods, and snails can be collected out of the ledge-forming unit visible near the top of the road cut on the north side of the old road. Examine some of the loose blocks that have tumbled down because the best material is frequently exposed by weathering on the blocks. To the south Sanpete Valley is beyond the hills in the foreground which are held up by the Green River Formation. High peaks of the Wasatch Plateau rise beyond Sanpete Valley. To the southwest, directly ahead, can be seen a flat upland surface sloping to the east. This is a high-level terrace adjusted to a high former position of the San Pitch River, prior to later erosion. **Continue on South on U.S. Highway 89** across alluvial fan built from hills and high points in the Cedar Hills to the west into a tributary of San Pitch River.

28.6 Junction of road to the east to the farming community of Milburn, in Sanpete Valley. Rest area to the west. A water gap is formed in the cuesta of Green River Formation to the east. The prominent cuesta, to the southeast at ten o'clock, is held up by the same resistant limestone unit that occurs just east of the railroad overpass by the last stop.

Prominent shoulders on the west-facing edge of the Wasatch Plateau to the east are held up by resistant sandstone beds beneath the Flagstaff Limestone in the upper part of the Tertiary and Cretaceous North Horn sequence. The plateau surface on top is in the Flagstaff Limestone, however, Sanpete Valley is in part a partial graben, a long linear down faulted structure, with a prominent fault along the western margin, much in the fashion of a hinged trap door. The road continues on to the south in a subsequent valley carved in the Green River Formation, on top of the resistant massive cuesta-forming limestone.

32.9 Additional outcrops of the tuffaceous Green River Formation can be seen in road cuts to the south. The hills to the west are the Cedar Hills and are capped by volcanic rocks of the Moroni Formation.

33.5 Green River Limestone exposures at the south end of the cuesta. From here the road drops to the floodplain of San Pitch River.

33.8 Cross San Pitch River.

34.0 Enter the Community of Fairview. Junction of U.S. Highway 89 with Utah State Highway 31 east to Huntington up Cottonwood Canyon, the V- shaped notch to the east. Much of the northern part of the community of Fairview is built on a broad alluvial fan (fig. 2.7) constructed out of Cottonwood Canyon.

35.1 Leaving Fairview. Green River beds are exposed in the base of the hills on the west of the valley. The Wasatch Monocline is well expressed in Green River and Flagstaff rocks in the Wasatch Plateau east of the valley (fig. 2.8). The western margin of the valley may be faulted.

40.9 Railroad Crossing. Green River rocks exposed in the base of the Cedar Mountains to the west. The Gunnison Plateau forms the skyline beyond the Cedar Mountains, with Price River-North Horn rocks exposed at its base and Green River

Figure 2.7. View northward across the toe of alluvial fans at the mouth of Cottonwood Creek to the gentle cuesta of the lower part of the Green River Formation, as seen from the north edge of the community of Fairview at approximately Mile 33.9. Westward dipping beds on the Wasatch Monocline form the hills in the distance at the right.

rocks on the skyline. The road rises to the south onto the broad alluvial fan produced by Pleasant Creek and associated streams draining the western front of the Wasatch Plateau.

40.6 Utah State Highway 116 Junction with U.S. Highway 89 in Mt. Pleasant, State Highway 116 leads westward to Moroni and Nephi.

42.1 Leaving Mount Pleasant, at junction of Spring City road, Utah State Highway 114, with U.S. Highway 89. Large boulders on the surface of the alluvial fan have been transported from Pleasant Creek as mud-flow rafted debris. Moroni is the community to the northwest at two o'clock. Type locality of the Moroni Formation is in the Cedar Hills, north of town.

43.4 Mount Nebo (fig. 2.1) can be seen on the skyline to the northwest at the southern terminus of the Wasatch Mountains. Salt Creek Canyon separates the southern Wasatch Mountains from the more southerly Gunnison Plateau.

Figure 2.8. View southeastward to Brigham's Chair, a prominent peak in Flagstaff Limestone, southeast of Spring City. Rocks in the lower part of the escarpment are Paleocene and Eocene formations dipping westward along the Wasatch Monocline, along the east end of Sanpete Valley.

47.7 Junction of Utah State Highway 30 to Chester and Wales with U.S. Highway 89. Continue on U.S. Highway 89. Green River rocks form road cuts immediately south of the junction and for the next few miles the road is on the stripped cuesta dip slope of Green River Limestone. The small community of Wales is at the base of the Gunnison Plateau to the west at the base of the alluvial apron. Beyond Wales, at the head of the alluvial fan at the mouth of the canyon, is Wales Gap. Here steeply dipping reddish Price River, Indianola, and Morrison beds (fig. 2.9) are exposed along the base of the mountains. Most of the escarpment is composed of steeply to gently folded conglomerate of the North Horn Formation, however, capped by Flagstaff Limestone. Colton beds form the upper recessive slope and are capped by Green River beds which form the skyline. Geology of the base of the plateau is complex.

53.3 Junction of Utah State Highway 11, with U.S. Highway 89. The State Highway leads northwest to Moroni, and east to Spring City through a gap in the

a.- GUNNISON PLATEAU

System	Group	Formation / Member	Thickness	Description
OLIGOCENE		Moroni Volcanics (upper Moroni = Laguna Springs, lower Moroni = Packard equiv)	2200	quartz latite welded tuff rhyolite tuffs blue & green volcanic sands, silts, cg white ash red siltstone
EOCENE		Crazy Hollow Fm	200	
EOCENE		Green River Fm	30—500	gray and green sh ls, oolite tuff
EOCENE		Colton Fm	300-800	varicolored sh, ss
EOCENE / PALEOCENE		Flagstaff Fm	720	pisolitic ls, gray sh, ss
PALEOCENE		North Horn Fm	0-3200	red, gray, green sh ss, cg, some ls 1660 at Wales
CRETACEOUS (Mont'n)		Price River Fm	50-1000	325 at Wales UNCONFORMITY
CRETACEOUS (Mont'n)		South Flat Formation	950	Castlegate equivalent UNCONFORMITY
CRETACEOUS (Coloradoan)	Indianola Group	Member 4	1430	cg, ss, sh, ls, coal gray qtzt cg
CRETACEOUS (Coloradoan)	Indianola Group	Member 3	0-1400	tan cg
CRETACEOUS (Coloradoan)	Indianola Group	Member 2	0-180	greenish ss
CRETACEOUS (Lower)	Indianola Group	Member I ("Lower Indianola")	1800—5600	mostly red and gray cg, with some sh, ss and fresh water ls zones Paleozoic ls and dolomite pebbles are abundant
JURASSIC	Arapien Shale	Twist Gulch Member	1300—1900	380' Summerville 215' Curtis 1040' Entrada equiv red sh, silt, grit
JURASSIC	Arapien Shale	Twelvemile Canyon Member	2700	Carmel equivalent gray ls and shale, some ss, gypsum, and salt *Pentacrinus Camptonectes Trigonia*

b.- SANPETE VALLEY

System	Group	Formation / Member	Thickness	Description
EOCENE		Crazy Hollow Fm	220	
EOCENE		Green River Fm	1015	*crocodiles, fish, turtles*
EOCENE		Colton Fm	580	
EOCENE / PALEO		Flagstaff Fm	350	
PALEO		North Horn Fm	1400	
CRETACEOUS (Mont)	Mesaverde Group	Price River Fm	250	
CRETACEOUS (Mont)	Mesaverde Group	Castlegate Ss	580	
CRETACEOUS	Mesaverde Group	Blackhawk Formation	1750	
CRETACEOUS (Coloradoan)	Indianola Group (in Sixmile Creek south of Manti)	Sixmile Canyon Formation	2800	gray and tan ss and cg, some coal
CRETACEOUS (Coloradoan)	Indianola Group (in Sixmile Creek south of Manti)	Funk Valley Formation	2250	tan ss, local cg, some marine shale
CRETACEOUS (Coloradoan)	Indianola Group (in Sixmile Creek south of Manti)	Allen Valley Shale	600—900	gray marine shale
CRETACEOUS (Coloradoan)	Indianola Group (in Sixmile Creek south of Manti)	Sanpete Formation	1350	tan and gray ss, cg
JURASSIC		Morrison? Fm	300—2000	
JURASSIC	Arapien Shale	Twist Gulch Member	1910	100' Summerville 230' Curtis 1580' Entrada equiv *Ostrea*
JURASSIC	Arapien Shale	Twelvemile Canyon Member	3500	Carmel equivalent

Figure 2.9. Stratigraphic sections of rocks exposed in the Gunnison Plateau and in Sanpete Valley along Route Segment 2. The Gunnison Plateau is to the west of the route and the Sanpete Valley section is that generally exposed along the base of the Wasatch Monocline on the east side of the valley (from Hintze, 1973).

Green River cuesta. Green River rocks are exposed in road cuts to the south, at the base of the west-dipping Green River Formation cuesta.

55.5 Old loading hoist. Blocks of oolitic Green River limestone were loaded onto the railroad here. The limestone was obtained in the large quarry visible to the east at the base of the monocline. The stone was used in construction of many older buildings throughout Sanpete Valley.

59.5 Junction of Utah State Highway 29 with U.S. Highway 89 in the southern part of Ephriam. The State Highway leads eastward over the Wasatch Plateau.

61.8 Low rolling hummocky topography at the base of the Wasatch Plateau to the east is carved on slumped Colton beds (fig. 2.9), stripped from the west-dipping surface of the Flagstaff Limestone. The low basal hills along the east side of the valley from here to south of Manti are residual hills of the slump. To the west, Price River and North Horn beds are exposed in the base of the Gunnison Plateau. The Flagstaff Limestone forms a prominent shoulder below the retreating Colton Formation that is capped in the high country by Green River beds.

63.6 Prominent quarry in Green River Formation ahead at eleven o'clock in the one from which much of the fossiliferous oolitic limestone for the Manti temple was obtained. Flagstaff Limestone forms the prominent cuesta at the front of the Wasatch Plateau and clearly shows the faulted monoclinal structure of the western margin of the plateau. Red North Horn rocks are exposed beneath the Flagstaff caprock in Manti Canyon, the major break in the cuesta south of the ridge with the block M.

65.3 Entering Manti at the northeast edge of the cemetery.

65.7 Small park at the west base of the Manti temple hill (fig. 2.10) in a broad S-curve in the highway, at the north end of town. Many of the older homes are built of Green River Limestone.

Figure 2.10. Mormon Temple at Manti is made of cream-colored colitic limestone which was quarried from the Green River beds in the northeastern edge of Manti.

66.9 Curve at the south end of Manti. Reddish colored hills to the southeast are slump masses of Colton which moved down the west-dipping Wasatch Monocline. The monocline is somewhat faulted, as shown by offsets in Flagstaff Limestone (fig. 2.11).

68.0 Small hydrogen sulphide-bearing springs to the east are in nearly vertical Cretaceous Funk Valley Formation of the Indianola Group. Distinctive prominent tan sandstones occur above, with some interbedded shale and minor coal in lower exposures.

68.8 Junction with the highway to Palisades State Park. The Palisades Road leads south through Allen Valley (fig. 2.9), carved in steeply folded shale, flanked on either side by Cretaceous sandstone hog-

Figure 2.11. View northeastward to fault- interrupted exposures of Flagstaff Limestone along the west side of the Wasatch Plateau, on the Wasatch Monocline.

backs. This is the type area of this shale which is a western tongue of Mancos Shale.

69.4 Crest of small alluvial fan. Vertical Cretaceous Sanpete Sandstone shows well in hills to the east. To the west reddish Arapien Shale is exposed in hills across the valley, capped by Green River Formation on the skyline to the west.

71.0 Crest of broad bend in highway. Toward the west massive conglomerates in Morrison(?), Price River, and North Horn Formations are visible beneath Green River beds.

72.0 Cross Six-mile Creek and Enter Sterling. To the west Arapien Shale forms the immediate hills and is capped by Green River Limestone. To the east, Flagstaff Limestone forms Black Mountain, the high prominent peak on the skyline on the Wasatch Plateau. Steep dips on the monocline can be seen directly ahead in Flagstaff and Green River Limestones south of the community of Sterling.

73.6 Gunnison Reservoir to the West. Directly ahead to the south in Arapien Valley and to the west in low hills are barren gypsiferous slopes of Arapien Shale.

73.9 Junction with Utah State Highway 137 to Mayfield. Continue ahead on U.S. Highway 89. To the south on the west flank of the Wasatch monocline Green River and Flagstaff Formations form cuestas above a valley carved on reddish Colton rocks between the two resistant units. The road turns toward the west beyond the sandstone and cuts through Arapien Shale exposures.

74.8 Bend in the road which is between Arapien Shale exposures at the south edge of a small reservoir. Directly ahead to the west are red conglomeratic beds which have been mapped as Jurassic Morrison but may be as young as North Horn Formation.

75.1 Cross Branch Line of D&RGW Railroad. Light-colored sandstones and soft pink shales to the north have been mapped as Morrison Formation, but are involved in complex structure and could be much younger.

75.3 Cross San Pitch River. Complexly folded, somewhat faulted Price River and North Horn beds form the low rounded hills to the west. Arapien Shale forms the barren hills to the south, while directly ahead the tan cliffs on the hills in midvalley are held up by Green River beds.

76.1 Vertical conglomerate in the North Horn-Price River on the north, beneath Flagstaff Limestone in marked angular unconformity (fig. 2.12).

76.6 Badly broken and fractured Flagstaff beds exposed in the road cuts. North-south trending Antelope Valley to the west is eroded in reddish Colton beds that are exposed some distance to the north of the road along the strike. Directly ahead faulted Green River beds form tan cliffs above greenish slopes of the lower part of the formation. To the southeast, barren Arapien Shale is well exposed in Arapien Valley, the type area.

77.6 The highway crosses a small east-

Figure 2.12. Unconformity between Prive River beds (P), North Horn beds (N), and overlying Flagstaff Limestone (F) at Mile 76.1. The relationships here have been variously interpreted as a result of flowage of Jurassic salt at depth or a result of pusles of mountain building during the Sevier orogeny.

west fault which drops the cliff-forming Green River Limestone down to near road level on the north but to the south it is some distance above road level. The road continues westward along the fault which has offset the west-dipping cuesta of Green River beds.

80.3 Junction with Utah State Highway 18 at the North Edge of Gunnison. The State Highway leads north and joins U.S. Highway 91 at Levan. Turn south toward downtown Gunnison, past the city park, on U.S. Highway 89 (fig. 2.13).

81.8 Leaving Gunnison and Entering Centerfield, opposite the Gunnison Valley Elementary School. Mt. Musinia, elevation 10,986 feet, is the high white nipple-shaped peak on the Wasatch Plateau skyline to the southeast. It is held up by Flagstaff Limestone which is exposed in the westward- dipping Wasatch monocline beyond the barren Arapien Shale hills. The Valley Mountains are to the west.

83.5 Abandoned sugar processing plant on the east. Barren Arapien Shale forms the low hills at the east edge of the valley, at the base of the Wasatch Plateau. The Fish Lake Plateau is the high volcanic-covered area east of the valley, 20 miles to the south.

87.5 Secondary road toward the west leads across the valley into the open-pit salt mine area in Arapien Shale. Another good road leads south from the mines to Redmond, the small community ahead at about two o'clock.

88.5 Junction Utah Highway 256 with U.S. Highway 89. The State Highway leads southwest to Redmond. Continue on U.S. Highway 89. The barren low hills across the valley to the west expose salt-bearing Arapien Shale. Directly ahead, down the highway, are barren exposures of gypsiferous Arapien Shale. Massive beds of gypsum have been mined near Sigurd where the gypsum is processed into plasterboard. Steeply dipping Green River, Flagstaff, and Colton beds are exposed in fault-broken cuestas immediately east of the road. Barren gray and pinkish exposures beyond the Tertiary cuestas are of Arapien Shale. The plateau skyline is capped in part by tuffaceous Late Tertiary volcanic rocks.

91.6 Small water gap through the Green River cuesta to the southeast shows Arapien Shale beneath slumped late Tertiary volcanic rocks along the plateau flank. Toward the southwest reddish rocks begin to appear in the Flagstaff Formation. These red units are traceable toward the south and grade into the Wasatch Formation exposed in the Bryce Canyon area. They are covered by volcanic rocks in the intervening Fish Lake and Sevier Plateaus. Far in the distance to the southwest the high peaks of the Tushar Range can be seen on the skyline.

93.5 Junction of Utah State Highway 256, south from Redmond, with U.S. Highway 89.

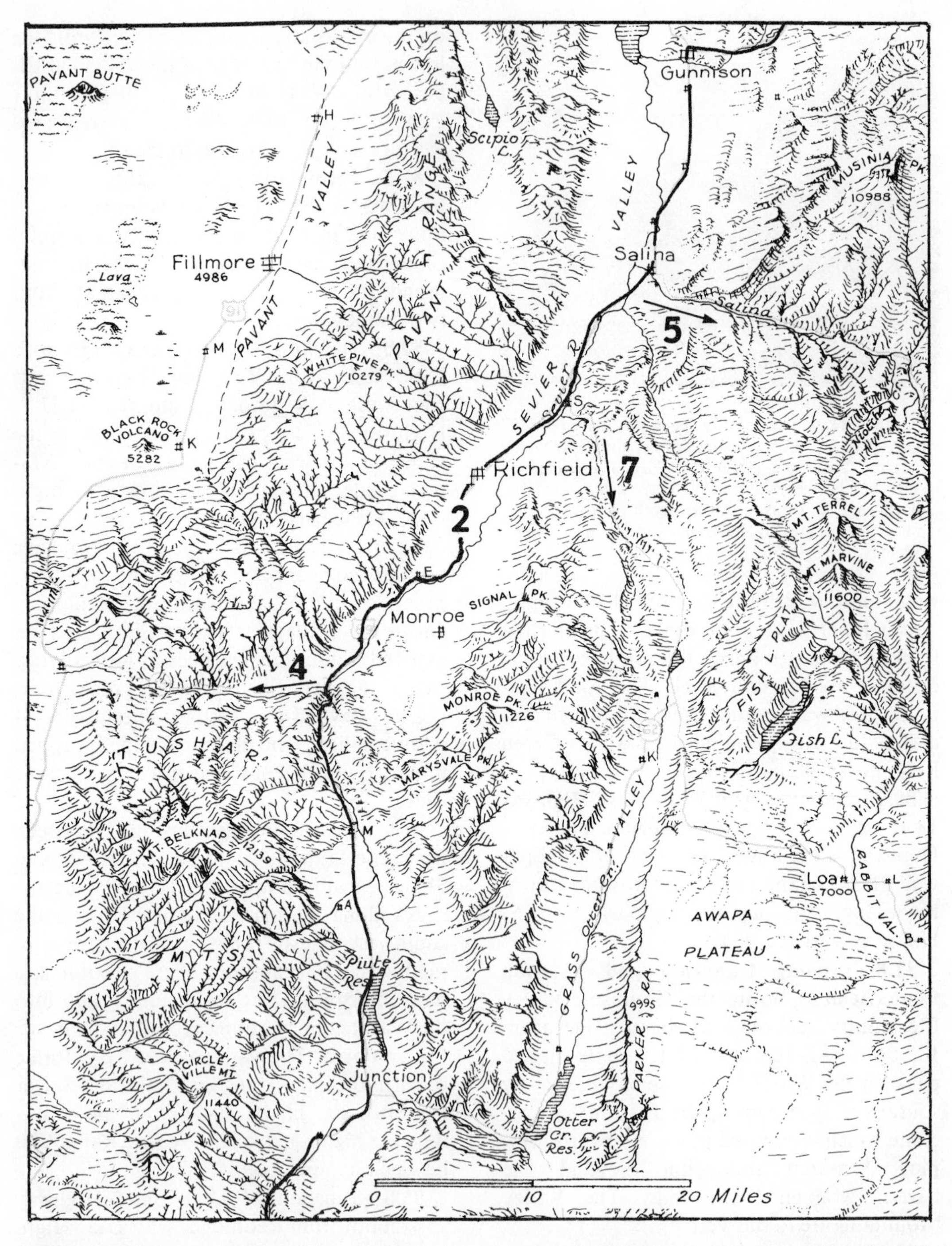

Figure 2.13. Index map of the middle part of the route of Segment 2. Route 5 joins Route 2 at Salina. Route 7 leads eastward from Segment 2 near Sigurd and Route 4 leads westward from near Joseph, west of Monroe.

94.6 Junction with Utah State Highway 4 in Center of Salina. Utah State Highway 4 leads toward Salina Canyon and junction with Interstate 70. See Guide Segment 5 for a log of that route to the east.

95.5 Leaving Salina at a broad bend toward the southwest. Toward the south, the prominent, high conical, barren light gray hills are in Arapien Shale, capped by purplish gray Tertiary volcanic rocks (fig. 2.14). The gentle eastward dip of the Pavant Mountains shows toward the southwest where alternate striped light gray and red Paleocene beds are exposed along the west side of the valley.

Figure 2.14. Conical hills of Arapien Shale are capped by remnants of Tertiary volcanic rocks at the northwestern edge of the Fish Lake Plateau, as seen from approximately Mile 96.0.

96.9 Cross Sevier River. The Sevier River drains northward from the Fish Lake Plateau and the Tuschar Range, which are toward the south, ultimately ending up in Sevier playa to the west of Delta. Hills to the east are capped by purplish andesitic and basaltic rocks apparently veneering a pediment cut across the poorly exposed Arapien Shale.

99.1 Road junction west to Aurora at the Fullers Earth-Bentonite-Rock Dust Plant on the east side of the road. Beyond the plant, excavations in gypsiferous Arapien Shale show well in the western base of the Fish Lake Plateau. These are the mines from which gypsum is obtained that is processed to plasterboard near Sigurd, ahead. The Arapien Shale is overlain unconformably by volcanic rocks of the Fish Lake Plateau to the southeast.

Red beds to the southwest of Aurora are in Boullion Canyon-Dry Hollow volcanics (fig. 2.15) in the youngest part and Cedar Breaks Formation (fig. 2.16) in the older part. The prominent striped sandstone beds to the southeast grade into the marly units of the Paleocene deposits that are now eroded to form Bryce Canyon and Cedar Breaks. These units are equivalent to the Flagstaff Limestone exposed to the northeast in the Wasatch Plateau. Prominent tan and light green rocks above the reddish beds to the southwest are some of the southwesternmost exposures of the Green River Formation in the state.

103.7 Junction Utah State Highway 24 with U.S. Highway 89. State Highway 24 leads east toward Fish Lake Plateau through the town of Sigurd, which is just to the southeast of the junction. A short distance beyond the town can be seen two of the large gypsum processing plants making plasterboard out of the gypsum mined in the Arapien Shale in areas to the east and northeast. Utah State Highway 24 leads eastward into Capitol Reef National Park. For a description of that route see Geologic Guide Segment 7.

105.3 Sigurd Freight Station, along the Denver and Rio Grande Western Railroad. The plasterboard plants show as a plume of white gypsum powder and steam in areas to the east, against the base of the Arapien Hills. To the south, Sevier Plateau volcanic mass at the western edge of the Fish Lake Plateau forms purple timber-covered bluffs

PAVANT RANGE

a.- SOUTH : KANOSH - COVE FORT AREA

System	Series	Formation	Thickness	Notes
Q		Pavant Basalt	0-200	
Q		Sevier River Fm	90	
MIDDLE TERTIARY		Bullion Canyon Dry Hollow volcanics	2560	
K		North Horn Fm	140	
K		Price River Cg	850	UNCONFORMITY
JUR		Navajo Sandstone	1740	
TRIASSIC		Chinle Fm	270	
TRIASSIC		Shinarump Cg	430	*petrified wood* sandstone grit
TRIASSIC		Moenkopi Fm	1050	*Pentacrinus* *Meekoceras*
PERMIAN		Kaibab – Toroweap Limestone	1190	*Dictyoclostus ivesi* *Huestedia* *Derbyia*
PERMIAN		Talisman Qtzt	30-290	
PERMIAN		Pakoon Dolomite	175	*fusilinids*
IP		Callville Ls	200	*Composita*
M		Redwall Ls	900	*Endothyra* *Euomphalus*
DEV		Cove Fort Qtzt	80	
DEV		Guilmette Fm	570	*Ceonites (coral)*
DEV		Simonson Dolo	240	
DEV		Sevy Dolomite	670	
S		Laketown-Fish Haven Dolomite	1000	*Halysites* *Streptelasma*
ORD		Eureka-Swan Peak Qt	170	
ORD		Pogonip Ls	1110	*Orthambonites* *Lachnostoma*
CAMBRIAN	Upper	Upper Cambrian interval not measured because of faulting		
CAMBRIAN	Middle	Cole Canyon - Bluebird - Herkimer Fms undiff	1150	
CAMBRIAN	Middle	Dagmar Dolomite	104	light gray
CAMBRIAN	Middle	Teutonic Ls	425	
CAMBRIAN	Middle	Ophir Formation	420	*Chancia ebdome*
CAMBRIAN	Lower	Tintic Quartzite	1300+ base not exposed	

b.- CENTRAL : FILLMORE - RICHFIELD AREA

System	Series	Formation	Thickness	Notes
Q		Axtell Fm	0-100	
MO		Dry Hollow volc	0-200	tuff, flows 24.7 M.Y. K-Ar
Ø		Bald Knoll Fm	475	*Helix-snail*
EO		Crazy Hollow Fm	260	ss, sh, siltstone
EO		Green River Shale	50-700	*ganoid fish scales*
PALEOCENE		Flagstaff Fm (equals "Wasatch" of Pink Cliffs at Bryce)	1900	*Helix* *Viviparus*
PALEOCENE – UPPER CRETACEOUS		North Horn Fm	3300±	
UPPER CRETACEOUS		Price River Fm	0-2600	UNCONFORMITY
TR – J		Navajo Ss	2000+	light red cross-bedded ss; THRUST
D		Sevy Dolomite	160	
S		Laketown – Fish Haven Dolomite	1200	
ORDOVICIAN		Swan Pk Quartzite	170	
ORDOVICIAN		Kanosh Shale	315	*Orthambonites*
ORDOVICIAN		Pogonip Ls	1120	
CAMBRIAN	Upper	Ajax Dolomite	700	
CAMBRIAN	Upper	Opex Dolomite	390	*Tricrepicephalus*
CAMBRIAN	Middle	"Cole Canyon" Dolo?	1500 – 2000	*Eldoradia*
CAMBRIAN	Middle	"Bluebird" Dolomite?	170-300	
CAMBRIAN	Middle	"Herkimer" Ls?	100-275	white weathering
CAMBRIAN	Middle	"Dagmar" Dolomite?	80	
CAMBRIAN	Middle	"Teutonic" Ls?	525	
CAMBRIAN	Middle	Ophir Fm	350	
CAMBRIAN	Lower	Tintic Quartzite	3000+ base not exposed	

Figure 2.15. Stratigraphic section of rocks exposed in the Pavant Range to west of Segment 2 (from Hintze, 1973).

Figure 2.16. Tertiary rocks exposed in the eastern edge of the Pavant Range from near Venice. Redbeds form the dark exposures near the base of the escarpment and are fluvial deposits which are overlain by lighter colored lacustrine rocks of the Green River Formation.

reaching up to approximately 9,000 feet. Banded Paleocene and Eocene rocks show very well in exposures along the base of the Pavant Mountains to the west, with interfingering of the more deltaic redbed series with the tan lacustrian beds.

107.9 Junction of a side road south to Venice. The Sevier fault is one of the main north-south-trending faults at the western edge of the Colorado Plateau. The fault trace is along the east side of the valley against the base of the Sevier Plateau. U.S. Highway 89 more or less parallels the fault from here southward to the Mount Carmel Junction area, east of Zion Canyon.

111.5 Broad turn northeast of the town of Richfield. The redbeds and sandy beds of the upper North Horn and Flagstaff equivalents are seen beneath the volcanic cover toward the southwest in the Pavant Range west of the town of Richfield. The geologic structure here is plunging toward the south, beneath the Marysvale volcanic field. A thick pile of andesitic and tuffaceous agglomeratic volcanic rock is exposed in Glenwood Mountain, the base of which is along the Sevier Fault.

112.7 Road junction of Utah State Highway 119, east to Glenwood, with U.S. Highway 89 at the east edge of Richfield.

113.5 Right Angle Turn South on U.S. Highway 89 into Downtown Business Section of Richfield.

114.6 Leaving Richfield. Volcanic rocks blanket the uppermost sedimentary sequence in canyons to the west of Richfield. The peaks directly ahead to the south and the southeast are all part of the Fish Lake and Marysvale volcanic district, through which we will now pass for several miles.

118.7 Road junction east to Anabella. High craggy exposures of volcanic rocks form Signal Peak and Glenwood Mountain, on the Sevier Plateau skyline to the east (fig. 2.17). The lower limit of juniper along the hills to the east marks the trace of the Sevier Fault.

Figure 2.17. Western escarpment of the Sevier Plateau. Glenwood Mountain forms the high promontory on the skyline. The Sevier Fault is at the base of the escarpment, near the light colored tuffaceous unit in the volcanic sequence.

119.8 Junction Utah State Highway 188 South to Monroe with U.S. Highway 89. Continue ahead on U.S. Highway 89. Volcanic rocks to the west are purplish andesitic breccia and tuffaceous units. Tuffaceous units form the lighter colored exposures in inliers in the canyon bottoms.

121.0 Entering Elsinore. Volcanic rocks to the west in the Pavant Range are the same series that we've been seeing for some distance to the north. These are some of the northwesternmost parts of the Mount Belknap and Fish Lake volcanic rocks. Some of the tuffaceous volcanic units are workable for building stone and some of the older buildings in town are composed of the softer units.

124.1 Well-bedded volcanic rock series show to the south. To the north, the alluvial fan gravels have been worked for road metal and the brecciated appearance of even the fragments shows very well. At about Elsinore, the Sevier Valley swings westward in its headwaters, west of the Sevier Fault escarpment, and the valley narrows down rather abruptly.

126.9 Entering Joseph. Massive tuffaceous units show as light hoodoos beneath the purplish andesites across the Sevier Valley to the east. Equivalent breccia units are exposed along the narrows of the canyon to the southwest of town.

127.7 Junction of Utah State Highway 188 east to Monroe in the south part of town. Continue ahead on U.S. Highway 89.

128.3 Leaving Joseph. Some of the large, partially buried structures here are potato cellars. Sevier Valley has been a major potato producer for some time. The road continues over alluvial fan material derived from erosion of the hills toward the northwest. Terrace gravels veneer the alluvial fan.

129.9 Entering Sevier. Notice the excellent angular unconformity to the north (fig. 2.18) with terrace gravels overlying light-colored, eastward-tilted tuffs. Just beyond the curve, notice the house built of tuffaceous volcanic rocks.

Figure 2.18. View northward of eastward tilted tuffs in the Tertiary volcanic sequence which are overlain with pediment-veneering gravels in a spectacular angular unconformity. The exposures are north of Sevier at approximately Mile 129.0.

130.6 Schoolhouse built out of the same tuffaceous material, now painted white. To the south, massive castellate and cliff-forming tuffaceous breccias show well at the northern edge of the volcanic field.

131.2 Junction of Utah State Highway 4 West to Interstate Highway 15 at Cove Fort. For a description along Utah State Highway 4 see Geologic Guide Segment 4. Continue south on U.S. Highway 89 into the narrows along Sevier River.

131.9 Cross Sevier River. Massive brecciated volcanic rocks are exposed in cliffs and ledges on both sides.

132.5 Excellent exposures of steeply eastward-dipping agglomerate and tuffaceous units on both the north and the south sides of the canyon. For the next several miles some of the volcanic units will appear like clay. These have been altered by hot water associated with the volcanic activity.

133.5 Cross Sevier River through tuffaceous double roadcuts. Most of the breccias appear to be lahar deposits associated with the composite volcano at Mount Belknap to the southwest.

135.3 Coarse-grained green dike cutting through the massive pinnacle-forming volcanic rocks. The breccias here still dip toward the east off Mount Belknap center.

136.3 Massive lahar breccias show very well about mid-canyon height on the east side across the Sevier River (fig. 2.19) in the Antelope Range. These rocks are some of the very coarse mudflow deposits, apparently associated with a composite cone to the west. On the west side of the valley, some yellow staining is now visible in the volcanic rocks where hot water alteration has produced kaoline and limonite-stained brownish clay.

Figure 2.19. Coarse volcanic breccias exposed along the canyon of Sevier River in the Antelope Range. These coarse accumulations are thought to be mud flow deposits associated with the Mt. Belknap volcanic center to the west.

138.0 Big Rock Candy Mountain Resort Area (fig. 2.20). The area around the resort, as well as to the east, has been altered by hot water, breaking down the volcanic rocks into the rather colorful peculiar assemblage of

Figure 2.20. Big Rock Candy Moutain as seen from the north at approximately Mile 138. The light, brightly colored, iron-bearing clays are the result of breakdown of minerals in the volcanic rocks. This alteration is though to have been associated with minor intrusions cutting the volcanic sequence.

iron-bearing minerals seen here. To the southeast, the strongly pinnacled region is a small igneous intrusion cutting up into the volcanic sequence. This intrusion may have been responsible for the hot water which produced the alteration of Big Rock Candy Mountain.

139.3 Broad sweeping curve. About midway through the curve, to the south, contact of a granite intrusion with the volcanic rocks to the west shows as the entrenched area. Some of the margins of the intrusion have been prospected for minerals. Road cuts on to the southeast are in badly weathered granite associated with the intrusion.

141.2 Prominent cliff of low-grade obsidian on the bluff on the west side of the highway as we enter Marysvale Valley. High gravel-capped terraces rim both the east and the west side of the valley. The Sevier Fault is along the east side of the valley, against the juniper-covered hills. To the northeast, the criss-crossing roads (fig. 2.21) mark a uranium mining area which underwent intensive exploration in the 1950s.

Figure 2.21. View northeastward from Mile 142 into the Marysvale Mining District, an area which received considerable attention during the uranium boom in the 1950s, as seen from the ridge at the north edge of Marysvale.

143.5 Leaving Marysvale. The road climbs up through terrace gravels of the Sevier River. The terraces here are cut across tuffaceous volcanic units of the Marysvale complex. Jumbled volcanic rocks on the east side of the valley are in the Sevier Fault area. The Tushar Mountains are to the west.

144.2 Area to the southwest criss-crossed by mine roads is on Deer Trail Mountain in the Marysvale alunite district. The district was worked primarily during World War I and World War II. The highway continues to the south for the next several miles, across terrace gravels and the toes of alluvial fans built out from the flanks of the volcanic Tuschar Range to the west. Across the valley to the east volcanic rocks of the south end of Sevier Plateau form the skyline

149.4 Crest of alluvial fan. The deeply entrenched Sevier River Valley is to the east and has a relatively narrow flood plain. Some distance to the east of the flood plain, a broad bajada surface rises up to the base of the plateau. The escarpment along the plateau marks the trace of the Sevier Fault. Hills to the southeast appear as rotated blocks west of the main Sevier Fault.

151.3 Crest of alluvial fan. Piute Reservoir is visible to the southeast. Steeply tilted tuffaceous and agglomeratic units form the alternating cliff and slope zone east of the reservoir (fig. 2.22).

Figure 2.22. Eastward across the north end of Piute Reservoir from approximately Mile 151. The gently tilted, gray, ashy beds in the foreground are part of the Tertiary volcanic pile exposed in the Sevier Plateau to the east and in the Tushar Mountains to the west. Piute Reservoir is made by damming the Sevier River in a narrows along the canyon west of the Sevier Fault.

158.4 Junction of Utah State Highway 153 with U.S. 89 in front of the Piute County Courthouse in Junction. Volcanic rocks to the east show a steep north and northwestward dip. This may be related to dip off a major composite volcano situated between here and Antimony to the east.

160.5 Junction of Utah State Highway 62 East to Antimony with U.S. Highway 89. Rocks in the major volcanic pile to the east show well along the canyon. Continue south on U.S. Highway 89. Volcanic rocks of the Paunsaugunt. Plateau (fig. 2.23) now form the skyline to the east.

164.9 Enter Circleville after crossing Sevier River.

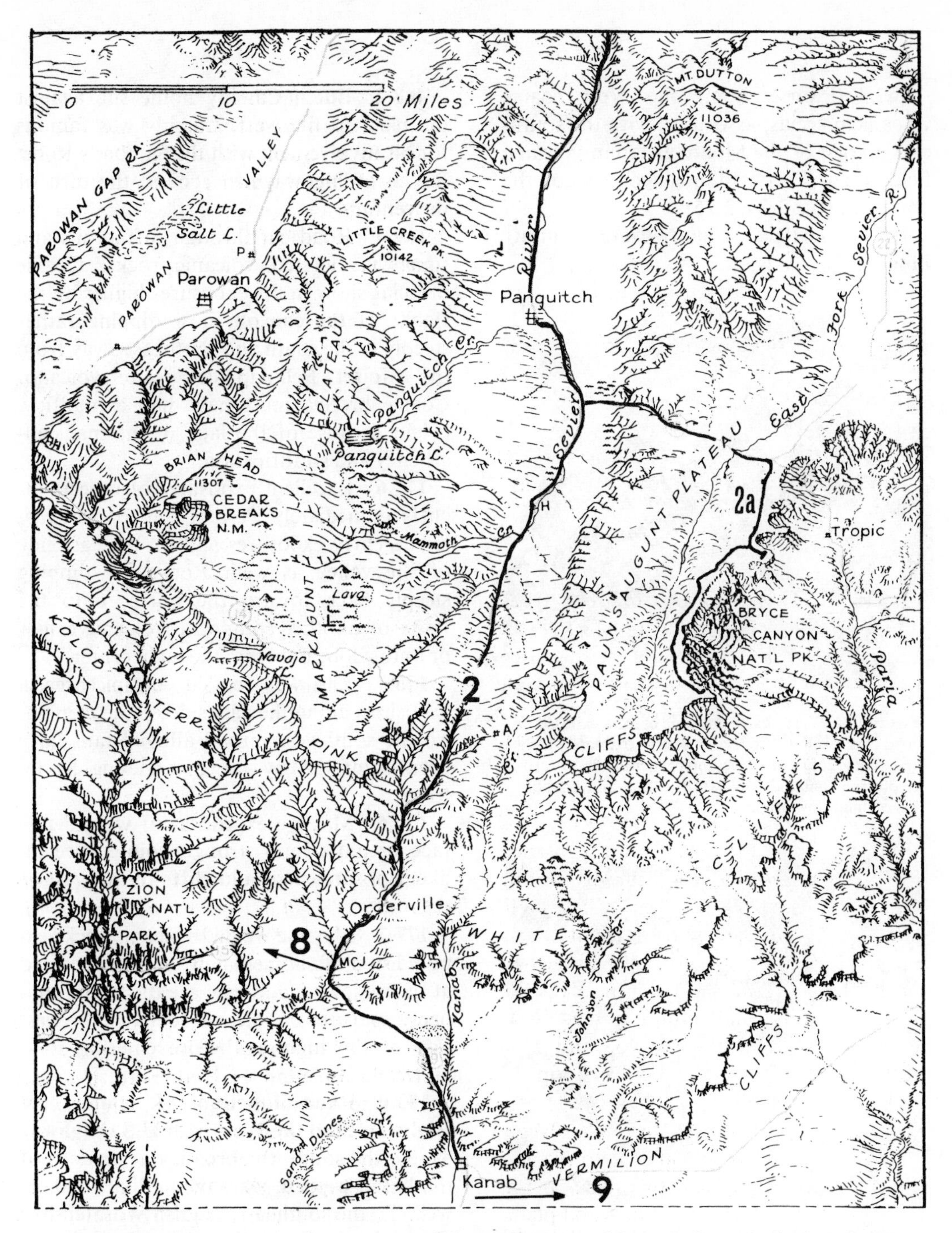

Figure 2.23. Index map showing the route of the southern part of Route 2. Route 2A leads to Bryce Canyon National Park and Route 8 leads westward through Zion National Park from near Mt. Carmel Junction. Route 9 continues southeastward across Glen Canyon and into the Navajo Indian Reservation.

167.3 Pioneer home site. Very coarse volcanic breccias, associated with the southeast margin of the Mount Belknap volcanic field, are well exposed in canyon cuts on the north side (fig. 2.24, 2.25).

167.6 Garfield County-Piute County Line.

Figure 2.24. Very coarse volcanic breccia exposed in road cuts south of Circleville, at approximately Mile 167. These coarse massive units generally show rather poor sorting and poor stratification and possibly have a mud flow origin.

Figure 2.25. Moderately well-bedded tuffaceous conglomeratic units in the lower part of the volcanic section, exposed south of the highway along the Sevier River, at approximately Mile 168.

168.2 Butch Cassidy home site against the bluff to the west. Cassidy was famous for his involvement with the Robber's Roost Bunch of highwaymen around the turn of the century.

169.4 Road cut through tuffaceous and somewhat altered volcanic rocks. Coarse breccias show in the exposures higher on the flanks of the canyon, now dipping rather steeply to the south and southeast away from the Mount Belknap composite volcano. Coarse breccia must have had a mudflow origin because of the lack of internal consistency and stratification.

171.6 Excellent exposures of breccia in-cliffs immediately west of the road in castellate cliffy exposures and on the east side of the valley as well. Rocks are dipping steeply back toward the north, onto the flank of another composite volcanic center or onto a lobe out from the main center.

175.1 Last well-bedded volcanic breccia exposures along the east side of the canyon. To the southwest, broad alluvial fans have been truncated by the meandering Sevier River.

176.6 Massive exposures of welded tuffaceous units form strongly jointed needle-like exposures and bold cliffs to the west of the road.

177.7 Massive welded volcanic rocks exposed in cliffs immediately west of the road at the major bend. These are the last prominent exposures of volcanic rocks on the west side of the highway before entering Panguitch Valley.

183.0 Road Junction of Utah State Highway 20 to the West with U.S. Highway 89. To the south, the broad, flat uplands of the basalt-capped Markagunt Plateau show well. To the southeast, reddish Wasatch-like Cedar Breaks Formation shows beneath basaltic flows at the south edge of the Fish Lake Plateau. Cedar Breaks National Monument and Bryce Canyon National Park are both carved in the reddish Tertiary beds where they are exposed high around the Markagunt and Paunsaugunt Plateau. The volcanic pile to the east shows a high angular castlelike feature on the skyline. These resistant cliff-forming units are probably the same as those that form the cliffs to the west. The Sevier Fault, downdropped on the west, is close against the base of the Fish Lake Plateau and the Tertiary reddish sedimentary rocks to the southeast across the valley, almost at the grasslands-juniper boundary. The highway continues ahead on volcanic debris swept out as alluvial fans along the west side of the Sevier Valley.

188.2 To the east, the reddish Tertiary Cedar Breaks Formation is overlain by a grayish slope zone and a massive cliff of the Brian Head volcanics (fig. 2.26). The volcanic unit forms the rim at Cedar Breaks, which is on the plateau margin to the southwest of us.

Figure 2.26. Eastward across Sevier River Valley southeast of Panguitch to the western face of the Fish Lake Plateau. The plateau is capped by gray cliff-forming Brian Head volcanic rocks that occur above the pink and light-colored Paleocene Cedar Breaks Formation. The Sevier Fault, which is the structural western boundary of the plateau, occurs in the middle distance at the base of the tree-covered slopes. Prominent terraces along the Sevier River show adjustment of the stream to a varying gradient downstream.

193.3 South End of the Business District in Panguitch. Turn east on U.S. Highway 89.

194.7 The road climbs off the flood plain of the Sevier River onto terrace gravels and alluvial fan material swept, in large part, from the plateau to the west. Stratification and internal structure of terrace gravels show well in road cuts around the point to the south.

199.1 Perched water table shows on the east side of the valley where a patch of grass marks emergence of water from the terrace. The road continues through terrace gravels which were deposited on a surface adjusted to a position of the Sevier River 50-60 feet above its present level. To the south and east reddish units of the Cedar Breaks Formaton (fig. 2.27) form pinkish exposures, cliffs, and some hoodoos. The somber, dark gray rocks in front, along the base of the hills are basalt, lowered along the Sevier Fault.

200.3 Junction of Utah State Highway 12 Access Road to Bryce Canyon National Park with U.S. Highway 89. At the mouth of the canyon, east along the highway, faulted basalts show very well downdropped against the older pinkish Bryce Canyon unit. For geologic guide to Utah State Highway 12 see Segment 2-A.

205.4 Cross-bedded terrace siltstone and gravel overlain by a basalt flow (fig. 2.28). The basalt is coming from one of the small volcanoes from the Markagunt Plateau in the vicinity of Panguitch Lake, off to the southwest, and has flowed down the east slope of the plateau into the Sevier River drainage.

205.9 Double road cuts through the basalt flow which is resting on terrace gravels and lacustrine (?) silt. The road continues parallel to the flow beyond the road cuts. The elongate ridge extending out from the plateau on the west is probably an inverted

BRYCE CANYON - KANAB AREA

System	Unit	Thickness	Remarks
TRIASSIC	Moenave Fm: Springdale Ss M	60	
	Moenave Fm: Whitmore Pt M	0-50	
	Moenave Fm: Dinosaur Can M	80-140	*Seminotus* (fish)
	Chinle Fm: Petrified Forest M	400	
	Chinle Fm: Shinarump M	0-160	
	Moenkopi Fm: upper red; Shnabkaib Member - 200'; middle red; Virgin Limestone Member - 50'; lower red	700—2000	bedded gypsum; *Tirolites*
PERMIAN	Kaibab Ls	250	Paleozoic beds known from well data
	White Rim? Ss	200	
	Toroweap Fm	100—500	
	Hermit Shale	0-100	Organ Rock equiv.
	Queantoweap Sandstone	1000	Cedar Mesa equiv
PENN	Callville Ls	200—600	Hermosa equivalent
MISS	Redwall Ls: Horseshoe Mesa M	170-290	cherty
	Redwall Ls: Mooney Falls M	240	
	Redwall Ls: Thunder Spr M	210	
	Redwall Ls: Whitmore Wash	100	cherty
D	Ouray Limestone M	0-150	
	Temple Butte Ls	110-250	Elbert equivalent
CAMBRIAN	Muav Limestone	1120—1230	
	Bright Angel Sh	330	
	Tapeats Ss	290—330	
LATE P€	Grand Canyon Series	820+	

System	Unit	Thickness	Remarks
PQ	basalt flows	0-300	locally present
	Sevier River Fm	0-60	
EO - OLIG - MIO	Volcanic rocks thickens from Panguitch area northword toward Marysvale area	0—several hundred	tuffaceaus andesites
	Brian Head Fm	0-300	
PALEO	Cedar Breaks Fm (=Wasatch Fm)	600	red and varicolored Bryce Cyn beds, fresh water deposit
	Pine Hollow Fm	0-400	
CRETACEOUS	Canaan Peak Fm	0-1000	
	Kaiparowits Formation	265—700	mostly bluish gray silts and sands, *fresh water fossils*
	Wahweap and Straight Cliffs Sandstones, undivided	535—1620	
	Tropic Shale = Mancos	625	marine beds; *Inoceramus Sciponoceras*
	Dakota Ss	175	coal beds
JURASSIC	Henrieville Sandstone	0-300	cut out by pre-Dakota erosion to the west
	Entrada Fm: Escalante Mbr	0-50	
	Entrada Fm: Cannonville M	0-230	
	Entrada Fm: Gunsight Butte	0-220	gypsum = "Curtis" of older reports
	Carmel Fm: Wiggler Wash M	0-60	
	Carmel Fm: Winsor Mbr	180-250	
	Carmel Fm: Paria Riv Gyp Mbr	60-160	*Pentacrinus abundant*
	Carmel Fm: Crystal Creek Mbr	0-180	
	Carmel Fm: Kolob Ls Mbr	70-230	
	Navajo Sandstone	1700—2000	
TRIASSIC	Tenney Canyon Ton of Kayenta Ss	0-120	
	Lamb Point Tongue of Navajo Ss	400	
	Kayenta Fm (main body)	300	

Figure 2.27. Stratigraphic section of rocks which occur in the Bryce Canyon-Kanab Area (from Hintze, 1973).

Figure 2.28. Weakly jointed basalt forms the canyon rim on a small tributary to the Sevier River at approximately Mile 205. Relatively thin basalt flows, such as the one exposed here, cap much of the Markagunt Plateau to the west.

valley. The basalt flowed down an ancient valley, filling it up, then armoring it. The enclosing softer material was then eroded away to leave the basalt-armored ridge. The long linear escarpment on the east side of the valley is a faultline scarp along the Sevier Fault. The plateau to the west is protected by an armor of overlapping thin sheets of basalt flows.

208.3 Entering Hatch.

210.6 The highway drops onto a basalt flow that has armored the eastward-dipping slope. The basalt is interbedded with some white tuffaceous material which is exposed just beyond the road cuts and a bridge. The road continues in white tuffaceous marly beds of the Sevier River Formation (fig. 2.27).

212.9 Pinkish exposures in the white marly beds in the uppermost part of the Cedar Breaks or the lowermost part of the Brian Head Formation. The escarpment to the east, at the base of the Tertiary red cliffs, marks the trace of the Sevier Fault.

214.2 Cross Asay Creek. The streams are entrenched into the light pink and cream-colored Cedar Breaks Formation. Moderately heavy growths of Ponderosa pine grow on hillsides above the flood plain.

217.3 Garfield County—Kane County Line. The road continues for the next several miles through the Cedar Breaks Formation.

221.4 Junction with Utah State Highway 14 which Leads West to Cedar Breaks National Monument. Continue straight ahead on U.S. Highway 89. The highway is in the lower part of the Brian Head or the upper part of the Cedar Breaks Formation, downdropped west of the Sevier Fault. Rocks equivalent to the resistant units exposed near the highway cap the Paunsaugunt Plateau to the east.

226.3 Road Junction Utah State Highway 136. U.S. Highway 89 continues southward through pinkish upper Cedar Breaks Formation. The broad, flat valley here is related to a basalt dam which plugged the small creek south of us.

228.2 The basalt dam forms the major topographic break in the valley ahead. The highway used to wind down through the narrow gorge to the east, where the basalt now shows as rubble. With modern road construction methods, the highway route has now been cut down through the cliff-lined valley wall west of the original gorge.

228.9 Cross-bedded bluff sandstone marks the upper part of the Cretaceous beds. Cross over large fill. The basalt flow seen at Mile 228.2 now forms a shelf to the east. The canyon to the west is cut down into Cretaceous rocks of the Kaiparowits and Wahweap Formation. Hoodoos are carved in the Cedar Breaks Formation near the skyline to the west in an area sometimes called Little Bryce Canyon (fig. 2.29).

229.9 Landslide in Cretaceous shales in

Figure 2.29. Bryce Canyon-like exposures of the Cedar Breaks Formation in "Little Bryce Canyon" at approximately Mile 229. Hoodoos and columns are carved in the lacustrine deposits in a continuation of the Pink Cliffs in the headwaters of Virgin River.

the east side. The slope was stabilized, but the cut for the highway caused some problems.

230.7 Roadside Rest Area and landslide region. Springs issuing from the sandstone have mobilized the pinkish shales of the upper part of the Cretaceous series. Hummocky topography in the rest area is the result of landslide masses moving down into the canyon. The V-shaped profile of the canyon, in general, is typical of stream-eroded canyons, particularly in areas where there are alternating shale and sandstone units like in the Cretaceous here. These rocks are lithologically equivalent to the upper delta plain facies like that in the upper Price River and North Horn sequence exposed in Price Canyon or Salina Canyon.

233.8 Cross-bedded sandstone and shale exposed in canyon cuts and road cuts on either side. The canyon flood plain begins to widen a little as the steepness of the stream course begins to decrease. Across the small creek in the canyon bottom, a thick cliff-forming barrier island sandstone is now exposed on the west side (fig. 2.30) and is the top of the coal-bearing Cretaceous sequence in Wahweap-Straight Cliffs beds.

Figure 2.30. Relatively massive, cliff-forming, barrier island sandstone in the Cretaceous section on the west side of the canyon is in the upper part of the coal-bearing Cretaceous deposits. Cretaceous rocks, west of the road, have been downdropped against massive Navajo Sandstone east of the road.

235.3 Entering Glendale. Prominent massive sandstone cliffs east of Glendale are held up by the Navajo Sandstone, upfaulted on the east side of the Sevier Fault. The Cretaceous rocks, which are coal bearing, have been downdropped to the west.

236.3 Glendale. Coal dumps on the east mark abandoned mines in some of the fault blocks in Dakota Sandstone. Behind the blocks, somber gray Tropic Shale is typical of much of the Cretaceous marine sequence. Barrier island sandstone with overlying coal-bearing lagoonal beds are exposed in road cuts. The Cretaceous beds are faulted and slumped down against the Navajo Sandstone which forms the prominent massive cross-bedded cliff east of the road.

238.4 Prominent red sandstone is the lower Navajo Sandstone, capped by white cross-bedded Navajo Sandstone in the cliff. The flat, well-bedded rocks on the skyline are the Carmel Formation. Those across the canyon to the west are in the Straight Cliffs Formation in the coal-bearing Cretaceous section.

239.7 Cross the Virgin River in Orderville.

240.2 South edge of Orderville. Massive sandstone to the southeast is the Navajo Sandstone (fig. 2.31). The somber soft shale in the immediate vicinity in the road cuts is Mancos or Tropic Shale.

Figure 2.31 Massive Navajo Sandstone exposed in the White Cliffs southeast of Orderville, as seen from approximately Mile 241. Cretaceous rocks form the Juniper- and pinion-covered exposures in the foreground, and faulted against the massive Jurassic Sandstone. Carmel Formation forms the bedded unit on the top of the cliff.

241.2 Cross small drainage. View back to the west shows overlying coal-bearing Cretaceous rocks and Mancos Shale above the Dakota Sandstone.

242.1 Entering Mount Carmel. Carmel Formation is exposed immediately west of the north entrance into town. Deep brick red units are exposed in som of the canyons west of the south end of town.

243.1 Cross Muddy Creek. Massive gypsum of Carmel Formation exposed in road cuts and forming the shoulder on the west side of the bluff, above cyclic bedded massive white siltstone and dark gray mudstones which appear much like the "stone baby" beds in Goblin Valley and other areas to the east.

243.9 Argillaceous, thin, platy beds of Carmel Limestone well exposed in road cuts on the west. The entire valley margin surface is overlain by a caliche-cemented terrace gravel, adjusted to a former high position of the Virgin River.

244.4 Junction of Utah State Highway 15, which Leads West Through Zion Canyon National Park, with U.S. Highway 89 at Mt. Carmel Junction (fig. 2.32). On the northwest corner of the intersection, across the small gully, platy limestone of the Carmel Formation is abundantly fossiliferous, containing spectacular accumulations of star-shaped *Pentacrinus* columnals. For a guide along Utah State Highway 15 and Zion Canyon see Geologic Guide Segments 8 and 8A.

Figure 2.32. View eastward to the White Cliffs over Mt. Carmel Junction. The well-bedded units in the foreground are Carmel Limestone on the down-dropped block, west of the Sevier Fault. The same bedded rocks are exposed above the massive Navajo Sandstone, on the skyline, east of the fault.

244.8 Cross the Virgin River in the lower part of the Carmel Formation. Road rises southward off the Virgin River flood plain. Directly down the river, upper massive beds of the Navajo Sandstone show in the canyon bottom. The road climbs through the fossiliferous marine and partially nonmarine sequence of the lower Carmel Formation. The same units through which we are passing on the highway cap the high skyline to the east of the Sevier Fault and give some measure of the fault displacement.

248.1 Side Road Leading South to Coral Pink Sand Dunes State Park. View to the north shows the offset on the Sevier Fault and Cretaceous stratigraphic units west of the fault line escarpment. Recent pink windblown sand overlies reddish cross-bedded windblown Triassic Kayenta Sandstone. The trace of the Sevier Fault crosses U.S. Highway 89 near here.

250.4 View Area, overlook to the south is over the northward dipping Triassic and Jurassic rocks of the Vermilion Cliffs rim above Kanab. The Kaibab Plateu rises in the far distance. North of us the White Cliffs in Navajo Sandstone (fig. 2.23) are traceable for some distance to the east. From here, the road descends into Kanab Creek Canyon through windblown pink sandstone and sagebrush-juniper woods. Reddish sand in the area is obviously being derived from erosion from the underlying soft cross-bedded red Terry Canyon tongue of the Kayenta that underlies the white Navajo Sandstone.

253.9 Cross-bedded, arkosic-appearing sandstone in the lower part of the red Terry Canyon tongue of the Kayenta.

255.0 Small swimming pool pond and joint-controlled springs in the headwaters of this fork of Kanab Creek. Excellent cross-bedded lower Kayenta Sandstone exposed in cliffs both east and west of the road (fig. 2.34).

Figure 2.33. View northward from the view area at approximately Mile 250. Massive cliffs of Navajo Sandstone are part of White Cliffs which extend around the southern edge of the Markagunt and Paunsagunt Plateaus. Navajo Sandstone is capped by moderately well-bedded units in the Carmel Formation.

Figure 2.34. Cross-bedded lower Kayenta Sandstone exposed in cliffs west of the road at approximately Mile 255. Southeastward (leftward) dip on the crossbed sets indicate direction of transport from the northwest in the ancient dunes. The exposures here are along a small fault that controls springs which occur to the left of the photograph.

255.8 Moqui Caverns, a tourist-oriented development in upper cross-bedded sandstone of the lower Lamb Point Tongue of the Navajo Sandstone, beneath the thin reddish Terry Canyon Tongue of the Kayenta Formation. Silica mines around the corner are in the upper part of the cross-bedded Lamb Point Tongue of the Navajo Sandstone (fig. 2.35) in the east canyon wall.

Figure 2.35. Cross-bedded Lamb Point Tongue of the Navajo Sandstone at Mile 256 is overlain by moderately well-bedded Kayenta rocks. The disturbed stratification was possibly a result of quicksandlike adjustment in the top of the Navajo Sandstone beds.

257.7 Cross Kanab Creek. Sedimentary structures within the secondary silt-filling show well near both bridge abuttments. Upper beds of the brick red sandy Kayenta Formation underlie the cross- bedded, white, lower Lamb Point Tongue of the Navajo Sandstone.

258.9 State Truck Stop, Highway Entrance Station.

259.8 Dam on Kanb Creek. The dam has controlled and produced the flat flood plain above the spillover. Kanab Creek below is incised into recent sand which had filled up the old valley approximately to road level (fig. 2.36). Sandy red Moenave beds are exposed both east and west. Maroon upper part of the Chinle Formation and lowermost part of the Moenave Formation are exposed in canyon walls to the south.

261.0 Entering Kanab City Limits.

262.3 Junction of U.S. Highway 89 with Alternate U.S. 89 in Southeastern Kanab. U.S. Highway 89 leads east toward Page. Alternate U.S. Highway 89 leads south to the north rim of Grand Canyon, and rejoins U.S. Highway 89 at Bitter Springs, Arizona, south of Page. For a continuation of the route guide along U.S. Highway 89 east to Page and south toward the Grand Canyon see Guide Segment 9.

Figure 2.36. Southeastward along entrenched Kanab Creek to a segment of the Vermillion Cliffs beyond. Kanab Creek is cutting into relatively recent silt and sand which have filled an earlier excavated valley. Cliffs in the background are in the Trassic Moenave Formation.

Segment 2A

0.0 Junction of Utah State Highway 12 with U.S. Highway 89. Utah State Highway 12 leads east to **Bryce Canyon National Park.** The state highway crosses the Sevier River immediately east of the junction. The Sevier River is entrenched into floodplain and possible lacustrine deposits.

7.2 Red Canyon Indian Store. Terraces of the Sevier River show well toward the south. To the north the grayish purple hills of Black Mountain are composed of volcanic rocks which have been downdropped west of the Sevier Fault.

9.1 Cross drainage of Red Canyon, approximately on the trace of the Sevier Fault. The fault is well displayed to the north where dark gray volcanic rocks on the west have been downdropped against the reddish Wasatch or Cedar Breaks Formation on the east (fig. 2.37).

9.2 Enter Dixie National Forest at the mouth of Red Canyon. The canyon and its spectacular display of erosional columns and hoodoos is carved in the Wasatch Formation. The irregularly weathering outcrops are characteristic of the marly, lacustrine rocks of the Wasatch Formation.

9.8 Parking area for Pink Ledges Trail on the north. The road continues to climb up through a sparse Ponderosa pine forest in relatively barren reddish Wasatch Formation.

Figure 2.37. The Sevier Fault as exposed at the mouth of Red Canyon at approximately Mile 9.1 on Route 2A. Gray basalt, on the left, has been lowered along the fault against the column-forming Cedar Breaks Formation, on the right.

10.4 Entrance to Red Canyon Campground on the south.

11.2-3 Tunnels through fins in the Wasatch Formation.

13.8 Summit at the head of Red Canyon. Prominent cliff toward the north is **Casto Bluff** held up by volcanic rocks at the southern end of the Sevier Volcanic Field. Ponderosa pine blankets the upper part of the Wasatch Formation to the south around sagebrush covered Coyote Hollow. The road continues to the east over the upper surface of the Paunsaugunt Plateau.

17.2 Rest area on the north in Wasatch Formation.

18.3 Cross the East Fork of the Sevier River, immediately west of a gas station- service area complex. To the east Aquarius Plateau is on the skyline. The Wasatch Formation erodes to form the Pink Cliffs around the plateau above the grayish Kaiparowits Formation' and below the light gray Brian Head volcanics that cap the plateau. The Aquarius block has been uplifted east of the Paunsaugunt Fault.

19.9 Access road to Bryce Canyon airport on the north in Emery Valley. Living quarters for the airport crew are on the south of the highway.

20.4 Junction of Utah State Highway 22 with Utah State Highway 12. Utah State Highway 22 to the east leads off the plateau and toward Escalante and other communities. Utah State Highway 12 continues toward the south to Bryce Canyon National Park. Turn south toward Bryce Canyon National Park.

21.7 Rubys Inn on the west. The highway continues to climb toward the park across a flat rimmed with Ponderosa pine.

22.4 Entering Dixie National Forest.

23.1 Entering Bryce Canyon National Park.

23.4 Junction of side road to Fairyland Canyon 1 mile to the east. The canyon is carved in Wasatch Formation, as is all the scenic upland of Bryce Canyon. Continue south into the main part of the park through marly upper Wasatch Formation.

24.2 Entrance Station and Visitors Center. Exhibits, rest room facilities, and literature are available in the Visitors Center.

24.7 Junction of side road east to lodge, grocery store, and Sunrise Point Overlook. Sunrise Overlook parking area is near the canyon rim north of Queens Garden and is approximately 0.2 mile off the main road, north of the lodge, but south of the grocery store. A service station is west of the junction.

25.4 Junction of Side Road East to Sunset Point. The side road leads approximately 0.2 miles east to the parking area. This overlook is one of the most dramatic and most easily accessible in the park and is on the rim of The Amphitheater above the Queens Garden section of the park (fig. 2.38). Trails begin near the parking area and loop down into the intricately eroded Wasatch beds. Headward erosion of various tributaries to the Paria River has combined with a regular joint pattern and variation in ease of erosion of various parts of the Wasatch Formation to produce here an unusual display.

Figure 2.38. View southeastward from Sunrise Point in Bryce Canyon across the Queen's Garden Amphitheater to Inspiration Point. Columns within the Queen's Garden Amphitheater are controlled by jointing and differential erosion of the marl beds within the lacustrine Cedar Breaks Formation.

25.9 Junction of Side Road East to Inspiration Point, Paria View, and Bryce Point

(fig. 2.39). The access road leads to easily accessible viewpoints on the southern and southwestern rim of The Amphitheater. These overlooks offer spectacular views of erosion features in the Wasatch Formation in the Pink Cliffs. Inspiration Point parking area is 0.4 miles to the east.

Figure 2.39. View northward across the Gueen's Garden Amphitheater to Sunrise and Sunset Points from Bryce Point. Aquarius Plateau forms the skyline in the distance on the right. Complex jointing and irregular differential weathering within the Cedar Breaks Formation produces the spectacular scenery at Bryce Canyon.

27.4 Aspens in the Ponderosa pine forest suggest that we are at an elevation of approximately 8,000 feet. Exposures of Wasatch Formation in the distance to the northwest are south of Red Canyon and west of the airport.

29.8 Viewpoint on the south of the road overlooks the rim of the Pink Cliffs and The Grand Staircase toward the south. Other view points ahead provide similar distant views for the next few miles but do vary in scenery along the Pink Cliffs.

35.9 Natural Bridge Parking Area. The bridge is immediately south of the highway and is capped by a resistant limestone in the pinkish beds of the upper part of the Wasatch Formation.

36.8 Viewpoint. Fir trees now begin to appear with aspen and Ponderosa pine on the northern slopes. The road continues to climb essentially in the same light gray to pink upper part of the Wasatch Formation.

37.5 Viewpoint on the east. In the far distance to the east the Aquarius Plateau and Boulder Mountain rise on the skyline above gray exposures of Cretaceous and older rocks.

41.7 Parking Area and End of Road at Rainbow Point and Yovimpa Point. Rainbow Point, to the north, looks out over the length of the Pink Cliffs in the park, as well as northward over the Paunsaugunt Plateau in the middle distance and the Markagunt, Sevier, and Awapa Plateaus in the far distance. Yovimpa Point, to the south, looks out over The Grand Staircase and northern Arizona. The Kaibab Plateau shows as a great uparch to the south and Navajo Mountain to the east rises above the Lake Powell country. **Turn around and return** toward the main park area.

Segment 3

0.0 Interchange at University Avenue South of Provo on Interstate Highway 15. The interstate highway is constructed over prodelta lake clays of Lake Bonneville or, in some areas, over younger organic marsh deposits. Provo City is currently using the area to the east of the highway for sanitary landfill and has modified the local drainage. Southeast of the intersection the road passes through marshes that border Provo Bay on Utah Lake (fig. 3.1). Water to the east is provided, in large part, by springs located near the base of the Wasatch Mountains or at the toe of the Provo River delta where porous sand was deposited over clay as the delta built southward and southwestward.

Wasatch Mountains along the east border of the valley are composed in large part of upper Paleozoic limestone but Precambrian and Cambrian rocks are exposed as the tan and very light gray lower part of the face of the front range. Dark gray outcrops in the lower part of the mountain front are fault-lowered slivers of Mississippian limestone and dolomite which were dropped along the Wasatch Fault. The Wasatch Fault trace is near the base of the mountain.

1.8 Cross beneath overpass at the northern Springville interchange. Complexly folded Mississippian rocks are exposed to the east in Buckley Mountain as the light and dark gray ledges along the front. Lime kilns have been developed in these rocks and in fault slivers of the same units, to the north.

The broad embayment in the Wasatch Mountains, to the southeast, traces out a pronounced bend in the Wasatch Fault. East and southeast of Springville most rocks in the mountain are within the Oquirrh Formation. Manning Canyon Shale (fig. 3.2) forms a strike valley that separates the higher peaks of Oquirrh Formation from the older Mississippian and Cambrian rocks of the front range.

3.8 Springville Interchange Overpass. Springville, to the east, is named for the numerous springs which rim the community. Some of the springs are fault controlled and issue along fault lines where porous sand and gravel have been juxtaposed with impervious clay. Other springs in and near the city issue where the contact of porous sand and gravel on underlying clay is exposed near the front of Hobble Creek delta. Much of the town is on this delta built into Lake Bonneville from Hobble Creek Canyon, the deep notch in the Wasatch Mountains to the east.

5.4 Overpass over Union Pacific Railroad Track. Maple Mountain to the east, between Hobble Creek and Spanish Fork Canyons, has well-defined triangular facets along its western margin. These are interpreted as evidence for recurrent movement along the Wasatch Fault which occurs near

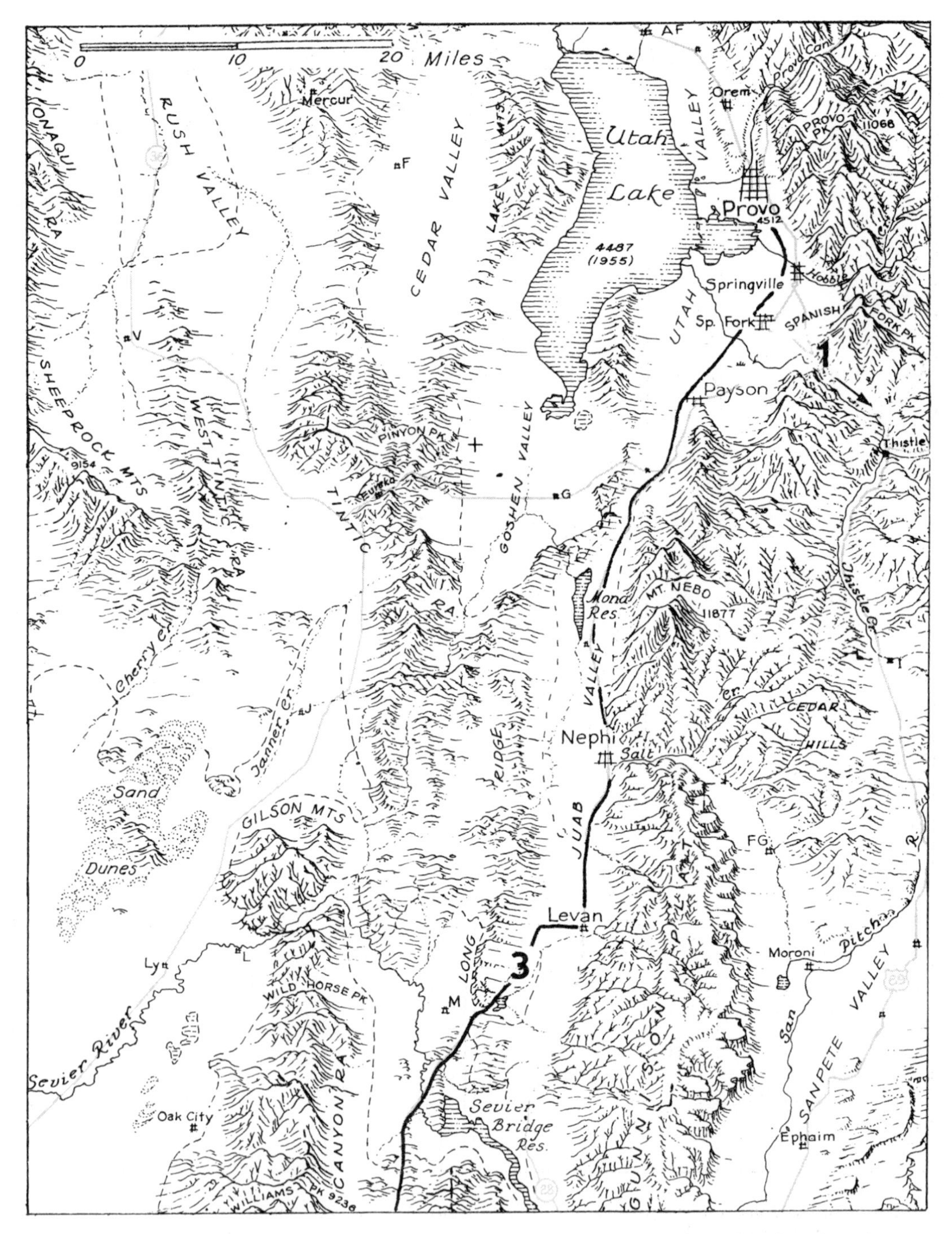

Figure 3.1. Index map of the northern part of Route 3. Route 1 diverges at Spanish Fork and heads up Spanish Fork Canyon toward the southeast. Segment 3 leads south through Juab Valley into the Scipio Area.

PROVO AREA - SPANISH FORK CANYON

System	Series	Formation	Unit	Thickness	Fossils / Notes
PERMIAN	Wolfcampian	Oquirrh Formation (not shown to scale)	Wolf-campian age Oquirrh 6200	19,000 (Hobble Creek Area) to 25,000 (Provo Canyon Area)	*Pseudoschwagerina*
PENNSYLVANIAN	Virg		Pole Canyon Member 2000		*Schwagerina*, *Waeringella*
	Mis		Lewiston Peak Member 2600		*Triticites*
	Desmoinsian		Cedar Fort Member 6200		*Wedekindellina*, *Fusilina*, *Chaetetes*
	A		Meadow Canyon Member 1100		*Profusilinella*
	M		Bridal Veil Falls Member 1245		*Endothyra* *brachiopods abundant*
MISSISSIPPIAN	Chesterian		Manning Canyon Shale	1650	*Lepidodendron*, *Rayonoceras*
	Chesterian / Meramecian		Great Blue Limestone	2700	*Archimedes*, *Fenestella*, *Apatognathus*
	Meramecian		Humbug Fm	930	*Fenestella*, *Polypora*
			Deseret Ls	850	*Gnathodus*
	Os / K		Gardison Ls	550	*Pseudopolygnathus*, *Syringopora*
D			Fitchville Fm	260	-white marker
CAMBRIAN	Middle		Maxfield Ls	600	*Kootenia, Spencia*, *Dolichometopus*
			Ophir Fm	250	*Glossopleura*
	L ?		Tintic Quartzite	1080	
PRE-Є			Mineral Fork Tillite	0-200	dolomitic clasts
			Big Cottonwood Fm	1000±	slate

System	Series	Formation	Thickness	Notes
Q		Lake Bonneville Group: Provo Fm, Bonneville, Alpine Fm	0-200	
Ø		Moroni Volcanics	0-500	= Laguna Springs & Packard of Tintic
EOCENE		Uinta Fm	0-500	
		Green River Fm	0-5000	thickens eastward rapidly
		Flagstaff Fm	0-600	*fresh water snails*
CRETACEOUS		North Horn Fm	0-2500	coarse cong at Red Narrows
		Price River Fm	700	ANGULAR UNCONFORMITY
		Indianola Fm	2600	red mudstone, sandy shale, cg
JURASSIC		Morrison Fm	1900?	*dinosaur bone*
		Summerville Fm	275	brown mudstone
		Curtis Fm	235	greenish gray shale
		Entrada Ss	1270	tan silty ss
		Twin Creek Ls (Carmel)	1570	*Pinna*, *Pentacrinus*
JURASSIC / TRIASSIC		Nugget Ss	1450	cross-bedded, Navajo equivalent
TRIASSIC		Ankareh Shale	1530	red shale (Chinle equiv), brownish red sandy shale (Moenkopi)
		Thaynes Fm	1340	ls, red shale, *Meekoceras*
		Woodside Shale	315	
PERMIAN	Wordian	Franson Memb of Park City Fm	600±	UNCONFORMITY, *Punctospirifer pulchra*
		Meade Peak Sh M Phosphoria Fm	175	phosphatic shale
	Leon	Grandeur Member Park City Fm	880	*Dictyoclostus ivesi*
	Leon / Wolfcampian	Diamond Creek Sandstone	835	
	Wolfcampian	Kirkman Ls	1200-	*Pseudoschwagerina*

Figure 3.2. Stratigraphic section of rocks exposed in the southern Wasatch Mountains in the Spanish Fork Canyon-Provo Area (from Hintze, 1973).

the Lake Bonneville shorelines at the mountain base.

5.8 Overpass over Denver and Rio Grande Western Railroad track. The delta of Spanish Fork River is well expressed to the east as the gentle rise and flat benchlands between the alkali marshes of the lake bottom near the freeway and the mountain front. This part of the delta formed during the Provo-level stillstand of the lake.

5.9 Interchange for Access Route to U.S. Highway 89 and 50-6 in northeastern Spanish Fork. U.S. Highway 89 leads through the Wasatch Mountains and then south parallel to Interstate Highway 15. For a description of the geology along U.S. Highway 89 see Geologic Guide Segment 2. U.S. Highway 50-6 leads southeastward toward Price and Grand Junction and for a description of that route see Geologic Guide Segment 1.

The U.S. Highways 89, 50-6 access road climbs up the foreset front of Spanish Fork delta and onto the flat topset plain of the Gilbertian-type delta. Interstate 15 continues southwestward over lake bottom clays.

6.7 Interchange at the north edge of Spanish Fork. Father Escalante and his companions were the first white men to enter Utah Valley. In September 1776 they came down Spanish Fork Canyon in their search for a route between Spanish colonies in New Mexico and California and camped three days near here with the friendly "Yuta" Indians.

Loafer Mountain, elevation 10,685 feet, is the high glaciated peak to the south. U-shaped valleys show well in the top, with cirques carved in Oquirrh Formation. Glaciers extended down to an elevation of approximately 9,000 feet.

West Mountain, to the west, is also composed in large part of Oquirrh Formation (fig. 3.2) that is part of a major thrust fault slice which apparently moved eastward 8 to 10 miles during the Sevier period of mountain building in the Cretaceous. Oquirrh rocks overrode Mississippian and older rocks that are exposed in the low hills south of the main range.

6.9 Cross over Spanish Fork River which has entrenched into its delta.

7.7 Overpass at the Benjamin interchange. Loafer Mountain to the southeast is flanked by a broad apron of pre-Lake Bonneville debris which has been dissected by young canyons. Dry Mountain, to the south, exposes metamorphic rocks along its lower western base and is capped by Mississippian rocks toward the east. The high flat uplands visible beyond between Dry Mountain and Loafer Mountain is the northern part of the Wasatch Plateau. The plateau is made up of relatively flat-lying Cretaceous and Tertiary rocks which were deposited over folded Paleozoic and Mesozoic formations.

The low ridge in the valley approximately one mile west of Interstate 15 is composed of Tertiary Salt Lake Formation and is bounded by a fault along its eastern margin. The fault is marked by a series of hot springs which extend northward for about 3 miles from the Benjamin Cemetery at the cluster of conifers on the ridge.

10.3 Payson Interchange, north edge of town. Interstate 15 continues over Lake Bonneville clays. A spit in the eastern part of town was built by Lake Bonneville where longshore currents were deflected by a hill of Tertiary volcanic rocks. Except for Dry Mountain, to the south, the other high peaks surrounding Utah Valley are held up by Oquirrh Formation, one of the thickest stratigraphic units in the state. The formation was deposited in a deep basin during Pennsylvanian and early Permian time.

14.5 Low rounded hills east of the overpass are on Precambrian gneiss and schist (fig. 3.3). Younger Precambrian and Paleozoic rocks form the ledges above. Lake Bonneville gravel and sand blanket the base of the mountains and are being quarried for construction materials. The highway now begins to climb up onto the toe of an alluvial fan built by Santaquin Creek.

Figure 3.3. View eastward from approximately Mile 14.5 near Santaquin of Precambrian (P), Cambrian (C), and Mississippian (M) rocks exposed along the western front of the Wasatch Mountains. The Wasatch Fault is near the base of the escarpment at the head of the alluvial fans.

16.6 Interchange of U.S. Highways 50 and 6 with Interstate 15. U.S. Highway 50-6 leads westward through the Tintic mining district toward Delta, Utah and Ely, Nevada. Godiva Mountain is the ridge on the skyline to the west and is in the center of the lead-silver producing Tintic mining district. Heyday of the district was about the turn of the century, but it has been recently reactivated by discovery of gold at depth northeast of Godiva Mountain.

17.7 Bridge over Santaquin Creek. Keigley Quarry to the north is in Cambrian, Devonian, and Mississipian limestone and dolomite and provides flux for the U.S. Steel Plant near Provo. Long Ridge is the low hills to the west and is a series of tilted fault blocks of Paleozoic and Tertiary rocks. Cretaceous and Tertiary North Horn Formation and volcanic rocks obscure Paleozoic formations in the hills west of Santaquin Canyon to the south. On the east side of Santaquin Canyon, however, Cambrian rocks are thrust over Oquirrh rocks.

19.1 Interchange southwest of Santaquin. Rounded low hills to the east are creep and landslide accumulations of Cretaceous-Tertiary North Horn Formation (fig. 3.4). Most movement must have been pre-Lake Bonneville because Bonneville-level sediments lap over the toe of the landslide debris. In the fields to the northwest the low elongate ridges are spits built out into Lake Bonneville at lower levels of the lake.

20.0 Utah County—Juab County Line at the crest of the hill. The county line is along a drainage divide at a bay bar of Lake Bonneville. The bay bar was formed at the Bonneville level of the lake and rests on a paleosol on Alpine level silt. These relationships are exposed in a railroad cut 0.3 of a mile west of the highway.

Road cuts to the south show the coarse bouldery debris of the landslides. Ahead of us to the south the reddish elongate hummocky tongue is a debris flow of North Horn material derived from exposures of the formation on the ridge to the east (fig. 3.5).

21.2 Crossing through road cuts on the toe of the debris flow tongue. Reddish clay matrix and boulders are typical. To the east Wasatch Fault at the base of the mountain front has offset all but the most recent sediments. In low morning light the scarp is particularly evident. Long Ridge makes up the western margin of the valley and is largely Paleozoic limestone but with a cap of Tertiary volcanic rocks which are related to those of the Tintic district on farther to the west.

23.5 Three cirques are present in steeply

MT. NEBO AREA, SOUTHERN WASATCH MTNS.

System	Series	Formation	Thickness	Remarks
PERMIAN	Wolfcampian	Oquirrh Fm	11,500 ±1000 (not shown to scale)	Bissell (1962 a) estimates the following thicknesses for the Oquirrh Fm here: 4000-Wolfcampian, 1000-Virgilian, 1500-Missourian, 3000-Desmoines, 1100-Atokan, 1000-Morrowan
PENNSYLVANIAN	Vir, Miss, Desmoin, Atokan, Mor			
MISSISSIPPIAN	Chesterian	Manning Canyon Shale	1200 ±200	slump areas, tree covered slopes
		Great Blue Ls	900 ±200	
	Meramecian	Humbug Fm	600	
		Deseret Ls	700	*crinoidal ls*
	Osage	Gardison Ls	340	*fossiliferous cherty ls*
	K	Fitchville Fm	190	*crinoid, coral, snail ghosts*
D	U	Opex Dolomite	290	oolites at base
CAMBRIAN	Middle	Cole Canyon Dol	460	interbedded light & dark gray dolomite
		Bluebird Dolomite	180	twiggy bodies
		Herkimer Ls	290	mottled, shaly ls
		Dagmar Dolomite	40	white, laminated dolo
		Teutonic Ls	330	mottled ls
		Ophir Fm	250	*Micromitra*
	L ?	Tintic Quartzite	930	
PRE €		Mutual Formation	370-1230	
		Farmington Canyon Complex?	—	

System	Series	Formation	Thickness	Remarks
TERTIARY		Salt Creek Fangl	0-100	
		Golden's Ranch Formation	600	green tuff & ss; buff tuff; andesite vol cong
		Flagstaff Fm	0-1200	*fresh water snails*
CRETACEOUS		North Horn-Price River Fms undifferentiated	1000±	massive red cg grit, sandstone, and algal ls
		Indianola Fm	1000±	algal ls; ss, cg; red shale
JURASSIC		Arapien Shale (much deformed by thrusting and folding): Unit 6	4000 ±1000	red sh, siltstone local salt, gypsum
		Unit 5		gray ls and sh; brown ss, cg
		Unit 4		gypsum, red shale
		Unit 3		gray splintery ls, ripplemarked gray siltstone and shale
		Unit 2		red & gray sh, gypsum
		Unit 1		shaly ls; gypsum; oolitic and bioclastic ls
		Triassic and older units lie on upper plate of Nebo Thrust Fault		
		Nugget Ss	600± est	salmon facies in upper thrust plate; buff and red in lower plate
TRIASSIC		Ankareh Fm	930	upper: sh, ss; middle: ss, cg, shale, siltstone
		Thaynes Fm	1350	red shale and siltstone; gray ls, shale, reddish sandstone
		Woodside Shale	1000	thinned by folding and thrusting to 100 ft in places
PERMIAN	Word	Franson Memb of Park City Fm	430	*bryozoans*
		Meade Peak Memb Phosphoria Fm	160	phosphatic shale
	Leo	Grandeur Member Park City Fm	510	*35' coquina of brachiopods*
		Diamond Creek Ss	330	slopeforming
	Wolf	Kirkman Ls	370	*Pseudoschwagerina uddeni*

Figure 3.4. Stratigraphic section of rocks exposed in the southern Wasatch Mountains in the Mt. Nebo Area. Paleozoic rocks can be seen on the western part of the range and Mesozoic and younger rocks on the eastern part, except right at the south end of Mt. Nebo where Triassic and Jurassic rocks are exposed in the frontal escarpment.

Figure 3.5. Tongue of landslide debris with a characteristic upper hummocky surface exposed east of the highway at Mile 21.2. The debris was derived from North Horn beds which are exposed in the ridge crest area.

dipping Oquirrh rocks (fig. 3.6) to the southeast near the crest of Mt. Nebo (elevation 11,928 feet). Glaciers extended down to approximately 9,000 feet, but below that the V-shaped valleys were stream-eroded and unmodified by glaciation.

28.3 North end of Mona. Mt. Nebo Reservoir west of town is along Current Creek and catches water from a series of springs that issue near the head of the reservoir. Water from the reservoir is used in southern Utah Valley. Purplish volcanic rocks on Long Ridge west of the reservoir bury Paleozoic limestone and dolomite.

29.6 South end of Mona. Hummocky topography of landslide or creep deposits on the lower part of the mountains to the east is produced by Manning Canyon Shale outcrops (fig. 3.7). The same Mississippian shale forms the pass that separates Dry Mountain from Mount Nebo and forms the shoulder on Mount Nebo. Pennsylvanian Oquirrh Formation makes up the upper part of Mount Nebo and Cambrian to Mississippian rocks make up the foothills to the northwest.

Figure 3.6. Mt. Nebo as seen from the northeast from approximately Mile 24.0. Three cirques are developed in the steeply dipping upper Paleozoic rocks near the crest of the mountain. Older Paleozoic rocks are exposed near the base and are cut by the Wasatch Fault which terminates the heads of most of the alluvial fans and the apron of debris at the base of the range.

Figure 3.7. The south end of Mt. Nebo and a major debris slide of hummocky material fron the Manning Canyon Shale as seen eastward from the south end of Mona at 19.6. Pennsylvanian Oquirrh Formation forms all the exposures above the landslide mass and is here steeply dipping in a drag fold of the Nebo overthrust fault.

Recent movement along the Wasatch Fault has produced a scarp at the base of the Wasatch Mountains. The fault has offset alluvial fans approximately 60 feet here at the base of the mountains.

32.3 Cross over the crest of an alluvial fan veneered with mud flow debris. Hummocky topography in the mouth of the canyon to the east is produced by broken Oquirrh rocks which were brecciated in the lower part of the overriding plate of the Nebo thrust fault. Fragmented Oquirrh rocks have been quarried for road metal along the front south of the canyons. From the quarry southward to near Nephi a younger fault scarp has displacement of up to 80 feet and marks the trace of the Wasatch Fault.

33.9 Side road west to Juab County-Nephi airport. The fault scarp shows as the break in alluvial fan profile at the base of the mountain to the east and southeast.

35.6 Northern City Limits of Nephi. The Nebo thrust fault is exposed in the reddish hills at the southwest base of Mount Nebo (fig. 3.8). The fault trace crops out at valley level almost directly below the block "J" where overturned and brecciated Oquirrh Formation occurs above overturned red Ankareh Formation (fig. 3.4). Park City and Phosphoria Formations are above the thrust fault on the ridgecrest on the skyline and are above overturned Navajo Sandstone which forms the light tan outcrop belt on the north side of a moderate canyon. Arapien Shale is exposed south of the Navajo Sandstone in the lower hills north of Salt Creek Canyon. Gypsum was mined from Arapien beds near the Nephi city dump until recently.

The Gunnison Plateau south of Salt Creek Canyon is not part of the overthrust fault system although gray Arapien Shale in the lower part of the hills was folded somewhat. Cliffs above the Arapien Shale are sandstone and conglomerate of the Cretaceous Indianola Formation (fig. 3.4) and were derived from erosion of the mountains produced by the Nebo thrusting and accompanying folding. Approximately equivalent and younger late Cretaceous beds bury the thrust fault toward the north and help date the thrusting as Cretaceous.

Figure 3.8. Nebo overthrust fault in Mesozoic and upper Paleozoic rocks at the southeast base of Mt. Nebo as seen northeastward from the north edge of Nephi. The fault is shown by a line and the overthrust direction by the arrow.

38.2 Intersection of Utah State Highway 132 with U.S. Highway 91 (and Temporary Interstate 15) in downtown **Nephi.** Interstate 15 is planned for construction east of town, but is temporarily routed along U.S. Highway 91 in various segments between Mona and Fillmore. Utah State Highway 132 leads eastward up Salt Creek Canyon and into Sanpete Valley and westward through the southern part of Long Ridge and around the north end of the Canyon Range toward Delta, Utah and Nevada.

39.7 Curve on U.S. Highway 91 south of Nephi. Low hills to the east are in gray Jurassic Arapien Shale. Cretaceous conglomerate and sandstone cap the high part of the Gunnison Plateau and are part of a thick clastic wedge deposited in the Rocky Mountain Geosyncline east of the mountains produced during the Sevier orogeny. Cretaceous rocks are reported to be approximately 20,000 feet thick 10 miles to the east.

43.2 Major curve on U.S. Highway 91 on the crest of Levan Ridge. The ridge is produced by merging of several alluvial fans from canyons off the Gunnison Plateau to the east. To the south the rounded irregular topography of the plateau margin is a function of the ease with which the Arapien Shale erodes. Cretaceous and Tertiary rocks cap the plateau. East dip of Tertiary Green River and Flagstaff rocks on Long Ridge' shows well by gentle eastern and steep western slopes to the west.

49.2 Levan. Junction of Utah State Highway 28 with U.S. Highway 91. The state highway leads south to Gunnison and U.S. Highway 89. Deep canyons of Chicken Creek and Pigeon Creek in the Gunnison Plateau to the east are carved in Arapien Shale. The same formation erodes to the rounded hills to the southeast of town and is overlain on the plateau crest by Cretaceous and Tertiary rocks. Beyond town U.S. Highway 91 swings westward across Juab Valley.

53.0 Major bend toward the south on the western side of Juab Valley. The highway swings parallel to the Union Pacific Railroad line. To the east we now get a long distance view of the Gunnison Plateau with soft-appearing rounded, somewhat barren to juniper-covered hills of Arapien Shale at the base. Cliffs east of Levan are in coarse clastic rocks of the Cretaceous Indianola Group. The top of the plateau southeast of Levan is capped by Tertiary Flagstaff, Colton, and Green River beds.

57.6 Chicken Creek Reservoir to the east. Flagstaff and Green River rocks form the cuestas to the west and dip eastward beneath Oligocene volcanic rocks.

58.3 Mills Junction Road and Chicken Creek Reservoir Dam. A variety of tuffaceous Oligocene volcanic rocks are exposed beyond the south dam and bridge abutment. White and tan lacustrine Green River and Flagstaff Limestones are exposed down the gorge to the west beneath the volcanic rocks and are overlain and underlain by redbeds.

60.2 Crest of hill in double roadcuts in fossiliferous green and tan Green River Formation. Similar beds are exposed to the north and south along the highway.

63.8 Bedded conglomerate of the Cretaceous-Tertiary North Horn Formation (fig. 3.9) is exposed east of the highway. These rocks are equivalent to those that occur near the top of the Gunnison Plateau to the east and along the eastern side of the Canyon Range to the west. They are part of the coarse debris fans swept eastward from the Cretaceous-age mountains produced at the site of the present-day Canyon Range and on westward into Nevada.

65.6 Cross Sevier River. The river meanders extensively here in the southern part of Mills Valley. The low gradient of Sevier River is in part produced by deposits of the stream which accumulated as a long delta into Lake Bonneville. The valley filled with fine-grained sediments during a high stand of the lake and is now being reexcavated following the last drop in lake level. The Sevier River drains around the northern end of the Canyon Range, across Sevier Desert, past Delta, and. into Sevier Playa, and has its origin on the Paunsaugunt and Markagunt Plateaus in the southern part of the state.

68.1 Roadside rest area is developed in Sevier River sediments associated with a high level of Lake Bonneville. Rounded conglomeratic ledges on either side of the road south of the rest area are North Horn Conglomerate. The lower ledges in the eastern part of

CANYON RANGE and NORTHERN PAVANT RANGE

System	Series / Group	Formation	Thickness	Notes
ORDOVICIAN		Watson Ranch Qtzt	315*	
	Pogonip Group	Kanosh Shale	370*	*Orthambonites* *Didymograptus*
		Juab Ls	70*	*Lachnostoma*
		Wahwah Ls	110*	
		Fillmore Ls	860*	*Trigonocera*
		House Ls	270*	*Hystricurus*
CAMBRIAN	Upper	Ajax? Dolomite	1000±	
		Opex? Ls	500±	
		Cole Canyon and Bluebird equivs.	1100±	
	Middle	Teutonic - Dagmar - Herkimer equivs.	720-900*	"*Ehmaniella*"
		Ophir Fm	250-400*	glauconite
	Lower	Tintic Quartzite	1500	pink to light gray thin even bedding, micaceous; lower part conglomeratic
		quartzite, tan and gray, medium to coarse grained conglomerate lenses common	1950	Woodward (1972) assignments:
LATE PRECAMBRIAN	Unnamed formations correlative with Sheeprock Series	shale, maroon, tan, micaceous	480	
		quartzite and conglomerate, light purple to red	1200	Mutual Fm
		sh, maroon, mica	75	
		sh, tan, micaceous	300	Inkom Fm
		sh, arenaceous	175	
		quartzite, tan to gray, med to coarse grained	1350	Caddy Canyon Quartzite
		qtzt and shale	335	Papoose Creek Formation
		ls, oolitic	60	
		qtzt and shale?	200	Blackrock Canyon Limestone
		ls, massive	140	
		ls, sandy, shaly	210	
		qtzt, greenish br	400	Upper Pocatello Formation ?
		shale with interbedded quartzite	640	
		qtzt, tan	440	

System	Series	Formation	Thickness	Notes
Q		Lake Bonneville Gp	0-300	silt, sand, gravel
TERTIARY	M	lava flow	0-50	tuff breccia
	Olig ?	Fool Creek Cg	0-1800	poorly sorted tan cg, many ls pebbles and cobbles
	Paleocene	Flagstaff Fm	1000±*	interlensing freshwater algal ls, ss, cg
		North Horn Formation	2600-3500	yellowish gray coarse ss, silt, cg; variegated shale with some freshwater algal marl; clastics come from the west
CRETACEOUS		Price River-Indianola equivalent?	3000*-12,500	gray to red, fine to coarse cg grading upwards into ss and shale with a few ls lenses
DEVONIAN		Simonson Dolomite	370	UNCONFORMITY; brown sugary dolomite
		Sevy Dolomite	1160	light gray weathering, fine grained
SIL		Laketown Dolomite	700±	dark gray, cherty
ORD		Fish Haven Dolomite	400±	
		Eureka Qtzt	155*	

Figure 3.9. Stratigraphic sections of rocks exposed in the Canyon Range and northern Pavant Range. The Canyon Range is west of the route and the Pavant Range is toward the east. A major unconformity is exposed at the base of the Cretaceous and the top of the Paleozoic sequence in both ranges' where clastic rocks derived from the Sevier orogeny rest on the older folded beds (from Hintze, 1973).

the Canyon Range are part of the same general conglomeratic section. These beds have been overridden by a thrust slice of Precambrian quartzite which is exposed as the lighter colored, well-bedded, rocks in cliffs in the upper half of the Canyon Range escarpment. The overthrust rocks are at the crest of the range from near Leamington Canyon, at the north end, southward approximately 20 miles to west of Scipio.

Oldest rocks exposed at the base of the range are the Ordovician Pogonip Group rocks which are gray to yellowish brown at the top and are overlain by the Ordovician Eureka-Swan Peak Quartzite which forms a thin, light tan cliffy bed. Somber gray Ordovician Fish Haven, Silurian Laketown, and lighter gray Devonian rocks occur immediately beneath the overthrust plate (fig. 3.10).

Figure 3.10. View westward from north of Scipio of the south edge of the Canyon Range. Distinctly bedded uppermost part of the ridge crest is held up by Precambrian rocks in an overthrust slide. Precambrian rocks here rest on Devonian beds beneath the fault. Rocks as old as lower Ordovician are exposed near the base of the escarpment toward the left.

71.7 Juab County—Millard County Line. Lower massive Cretaceous conglomerate forms cliffs below bedded over-thrust Precambrian quartzite in the Canyon Range to the west. North Horn Conglomerate and sandstone form the high country of the Valley Mountains to the southeast. Flagstaff Limestone caps the North Horn sequence along the eastern side of the range.

76.5 Junction of Utah State Highway 63 with U.S. Highway 91 and Interstate 15 north of **Scipio.** Directly south of town massive Cretaceous Price River Conglomerate caps the Pavant Range and rests unconformably over folded Ordovician, Silurian, and Devonian dolomite and limestone.

78.8 Beginning of major climb up into Scipio Pass through rather poorly exposed Ordovician Pogonip Group rocks.

82.0 Summit. Poorly fossiliferous Ordovician limestone exposed north of the summit area as well as in roadcuts on the southwest side. The highway now drops down the western side of the Pavant Range (fig. 3.12) across alluvial fans and valley fill.

The monument commemorates the epic journey of Father Escalante and Father Dominques and eight companions who traveled through this pass in October 1776 in their search for a route between Santa Fe, New Mexico and the new Spanish port of Monterey, California. They are the first white men known to have been through here.

90.9 Junction of Utah State Highway 26 with Interstate 15 and U.S. Highway 91 at the north edge of **Holden.** Utah State Highway 26 leads northwestward to Delta. Cambrian rocks are exposed on the western side of the Pavant Range and are unconformably overlain by Cretaceous Price River and North Horn beds at the range crest.

Pavant Butte, (fig. 3.12) the prominent peak in the valley to the west, is a volcano cinder cone. It is the site of a 1930s aborted attempt to generate electricity on a commercial scale utilizing windmills.

PAVANT RANGE
a. SOUTH : KANOSH - COVE FORT AREA

System	Formation	Thickness	Notes / Fossils
Q	Pavant Basalt	0-200	
Q	Sevier River Fm	90	
MIDDLE TERTIARY	Bullion Canyon Dry Hollow volcanics	2560	
K	North Horn Fm	140	
K	Price River Cg	850	
			UNCONFORMITY
JUR	Navajo Sandstone	1740	
TRIASSIC	Chinle Fm	270	*petrified wood*
TRIASSIC	Shinarump Cg	430	sandstone grit
TRIASSIC	Moenkopi Fm	1050	*Pentacrinus*, *Meekoceras*
PERMIAN	Kaibab – Toroweap Limestone	1190	*Dictyoclostus ivesi*, *Huestedia*, *Derbyia*
PERMIAN	Talisman Qtzt	30-290	
PERMIAN	Pakoon Dolomite	175	*fusilinids*
IP	Callville Ls	200	*Composita*
M	Redwall Ls	900	*Endothyra*, *Euomphalus*
DEV	Cove Fort Qtzt	80	
DEV	Guilmette Fm	570	*Ceonites (coral)*
DEV	Simonson Dolo	240	
DEV	Sevy Dolomite	670	
S	Laketown- Fish Haven Dolomite	1000	*Halysites*, *Streptelasma*
ORD	Eureka-Swan Peak Qt	170	
ORD	Pogonip Ls	1110	*Orthambonites*, *Lachnostoma*
CAMBRIAN Upper	Upper Cambrian interval not measured because of faulting		
CAMBRIAN Middle	Cole Canyon - Bluebird - Herkimer Fms undiff	1150	
CAMBRIAN Middle	Dagmar Dolomite	104	light gray
CAMBRIAN Middle	Teutonic Ls	425	
CAMBRIAN Middle	Ophir Formation	420	*Chancia ebdome*
CAMBRIAN Lower	Tintic Quartzite	1300+ base not exposed	

Figure 3.11. Stratigraphic section of rocks exposed in the Pavant Range in central Utah (from Hintze, 1973).

97.8 Cedar Mountain on the west is composed of eastward-dipping Tertiary volcanic rocks which are older than the volcanic rocks which have accumulated in the floor of the valley to the west.

99.6 Bridge of the Fillmore Interchange. Volcanic rocks of Black Rock Desert and Fillmore volcanic field occupy much of the lower parts of the valley to the west. Horseshoe volcano to the west is a cinder cone and is being quarried for construction materials. Tabernacle volcano to the southwest is the rounded hill that rises above the black basalt flows with which it and other volcanoes of the area are associated. Lavas erupted, in part, into Lake Bonneville and must have produced spectacular short-lived plumes. Other lavas predate the lake since some are reworked by lake currents, while still others are deposited on lake material and hence are younger than the lake.

Fillmore was the first capitol of Utah and the Capitol building was the first one constructed west of the Mississippi River. It housed the territorial legislature from 1855 until 1858 when the legislature adjourned to Salt Lake City.

103.1 Bridge over interstate at southern Fillmore interchange. Horseshoe volcano is to the west and Tabernacle volcano to the southwest.

107.9 Meadow and Kanosh Interchange and Junction with Utah State Highway 133. Additional volcanoes are visible to the southwest.

The white mound 5 miles to the west is known as White Mountain and is made largely of wind-blown gypsum derived from evaporation of local gypsiferous springs.

112.2 Directly west from here the low, tan, rounded hills are tufa cones associated with Hatton Hot Springs. The tufa has been quarried locally for building materials.

114.1 Rest area to the west is constructed

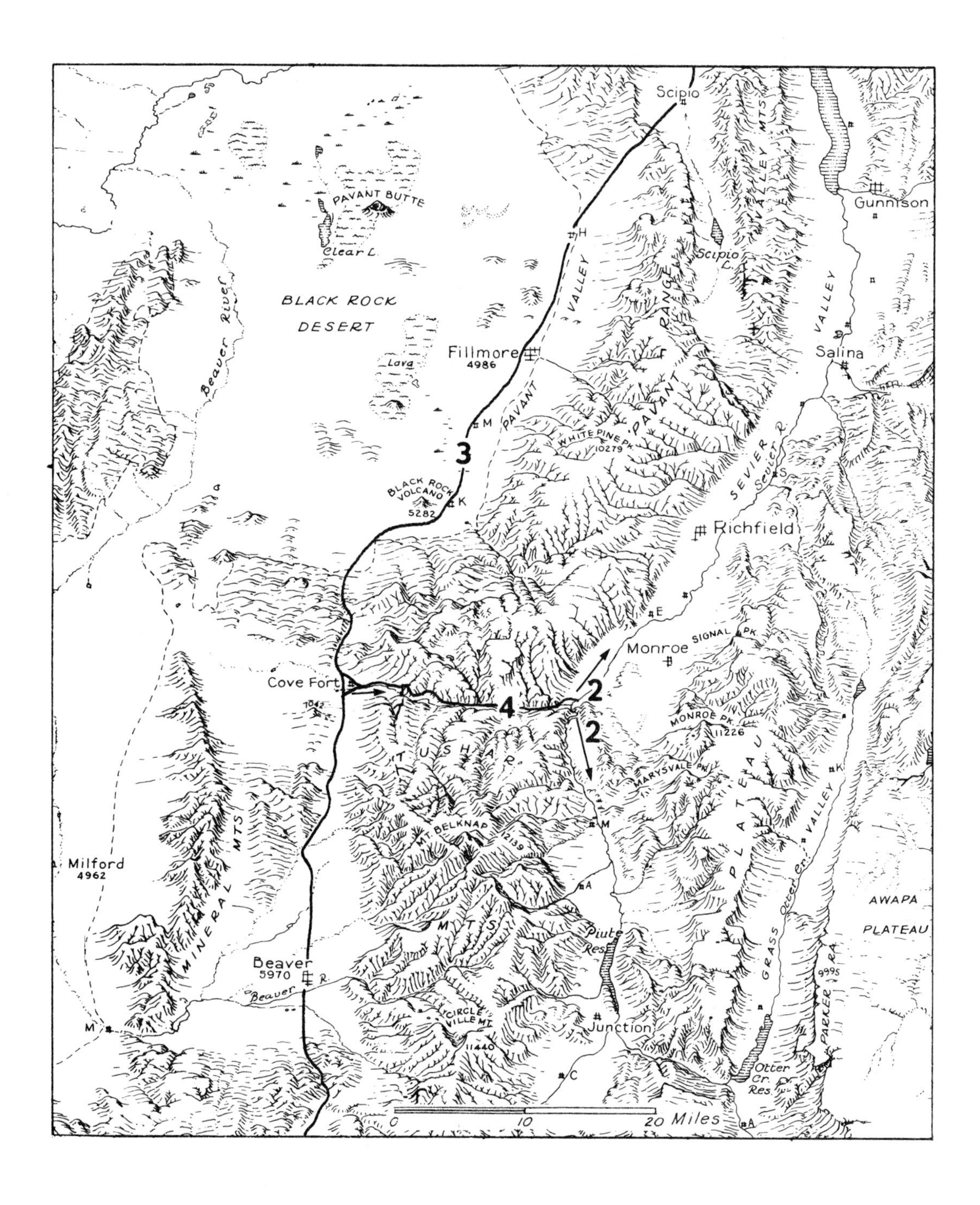

Figure 3.12. Index map of Route 3 in the central part of the area. Segment 4 branches eastward from Route 2 at Cove Fort interchanges and connects with Route 2 which is along U.S. Highway 89 in Sevier Valley.

on basalt associated with Black Rock volcano to the southeast. Upper levels of Lake Bonneville have been etched into the base of the cinder cone. Other volcanoes rise as pyramid-shaped peaks to the southwest.

119.3 Low ridges to the southeast of the road are steeply dipping Cambrian rocks. Those forming cockscombs closest to the road are Tintic Quartzite. The rounded hills beyond are of Cambrian limestone and shale or younger rocks, and all are included in an overthrust sheet which involves the southwestern spur of the Pavant Range.

120.0 South Kanosh exit. Cambrian rocks and young basalt dikes show immediately east and southeast of the interchange.

120.5 Major bend in the highway. Shallow terraces of highest levels of Lake Bonneville form the flat country on either side of the highway. The highway rises to the south through the Paleozoic section which is poorly exposed here in sagebrush and juniper-covered hills.

123.1 Broken and overturned Ordovician Pogonip Group rocks. Some beds have been so badly broken that coherent bedding has been destroyed, particularly in some of the more easily sheared argillaceous units.

124.3 Brecciated Ordovician, Silurian, and Devonian units are overturned in a dragfold on the lower part of the thrust sheet.

125.1 Completely overturned Mississippian and Pennsylvanian rocks are exposed around the northern margin of Dog Valley. Mississippian Redwall Limestone is exposed near the road. Dog Valley has only internal drainage and must be a large sinkhole in the fragmented carbonate rocks associated with the thrust fault. Well-bedded Pennsylvanian and Permian rocks east of Dog Valley in the Pavant Range are in normal right-side-up position beneath the fault.

128.4 Tertiary volcanic flows and pyroclastic rocks exposed in road cuts on either side of a rest area in the southbound lane. Volcanic mudflow or lahar breccia and conglomerate constitute much of the ashy massive to bedded rocks.

129.3 Cove Fort Interchange with Utah State Highway 4. For a description of the geology along this route past Cove Fort and along Interstate Highway 70 and Utah State Highway 4 to the east see Geologic Guide Segment 4. This route connects east to U.S. Highway 89 which, in turn, connects with Interstate 70 again in the Salina, Utah area. Cove Fort was built at the request of Brigham Young out of locally quarried basalt and purplish andesite or latite. The fort was completed in 1867. It now houses a small museum.

131.4 Massive basalt is exposed on either side of the freeway. Flows have been chopped into innumerable blocks by small faults. The lava is associated with the large volcano to the south, which is about on the line between Millard County and Beaver County.

132.1 Bridge at the Junction of Interstate Highways 70 and 15. Geology along Interstate 70 east of Cove Fort is described in Geology Guide Segment 4.

134.9 Cross beneath bridge of Sulfurdale interchange. Between Cove Fort and here the workings at Sulfurdale can be seen periodically to the east against the base of the Tushar Mountains. Elemental sulfur is disseminated through tuffaceous rhyolitic rocks in the open pit of the inactive mine.

139.4 Cross beneath bridge at Ranch exit. The highway is constructed over old alluvial fans that are probably pre-Bonneville. Tertiary Bullion Canyon and Dry Hollow Formations are exposed in the face of the Tushar Mountains to the east. Upper ledges are in the Dry Hollow Formation.

141.3 Road cuts through tuffaceous

light-colored Sevier River Formation in both the south and northbound lanes. To the south the road drops down through younger high terrace gravels.

144.4 Bridge at the Manderfield Exit. The highway is in terrace gravels and valley fill, but to the west granitic peaks of the Mineral Range rise to 9,100 feet. Granite forms the prominent serrated peaks of the skyline. Mt. Belknap (elevation 12,139) and Delano Peak (elevation 12,173) are the high points in the Tushar Range to the east and are part of a large Late Tertiary composite volcano (fig. 3.13).

Figure 3.13. View eastward of Mt. Belknap from the north edge of Beaver at approximately Mile 163. Mt. Belknap is the high point in the Tushar Range and is, in part, the remnant of an ancient stratocone volcano that makes up much of the deposits of the range.

146.3 High point on the road provides a view southward over Beaver Valley and of the prominent gravel-covered pediments or terraces that surround the valley. Terraces were apparently adjusted to a high outlet through the gap to the west at the south end of the Mineral Range.

152.9 Beaver Interchange and Junction with Utah State Highway 21. Utah State Highway 21 leads west to Minersville and Milford, around the southern end of the Mineral Range.

The Mineral Range is the site of the oldest lead-silver mining venture in the state. The Lincoln Mine, discovered in 1852, produced lead utilized for bullets by early settlers but silver in this lead made it too hard to be extensively utilized. Silver made Beaver County famous for a while. Higher terraces show well east of town, particularly near the block letter "B."

156.1 Bridge and the South Beaver interchange. East of the bridge a gravel quarry in the terrace veneer shows the internal structure of the deposit. The marshy area in the vicinity of the road is produced by water which is forced to surface because of a bedrock sill at the west end of the valley.

161.4 Road cuts through pink and green Tertiary Bullion Canyon volcanic rocks on the southwest side of the Tushar volcanic center.

163.5 Late Tertiary tuffaceous rocks and basalt exposed in road cuts.

164.0 Beaver County—Iron County Line. To the south the highway drops into the valley of Fremont Wash. Varicolored volcanic rocks are poorly exposed on either side of the valley and weather to produce a thick soil overgrown by sagebrush and juniper woodlands.

170.8 Junction of Utah State Highway 20 with Interstate 15. High peaks to the east and west of the north end of Parowan Valley are in younger Tertiary volcanic rocks.

177.6 Roadside rest area off the southbound lane. Reddish Wasatch or Cedar Breaks Formation is now exposed beneath the volcanic rocks along the base of the fault escarpment to the east. This red formation rises toward the south and forms the Pink Cliffs and much of the scenic area around Cedar Breaks National Monument (fig. 3.14) and the high rim of the Markagunt Plateau.

181.1 A gentle anticline is expressed at

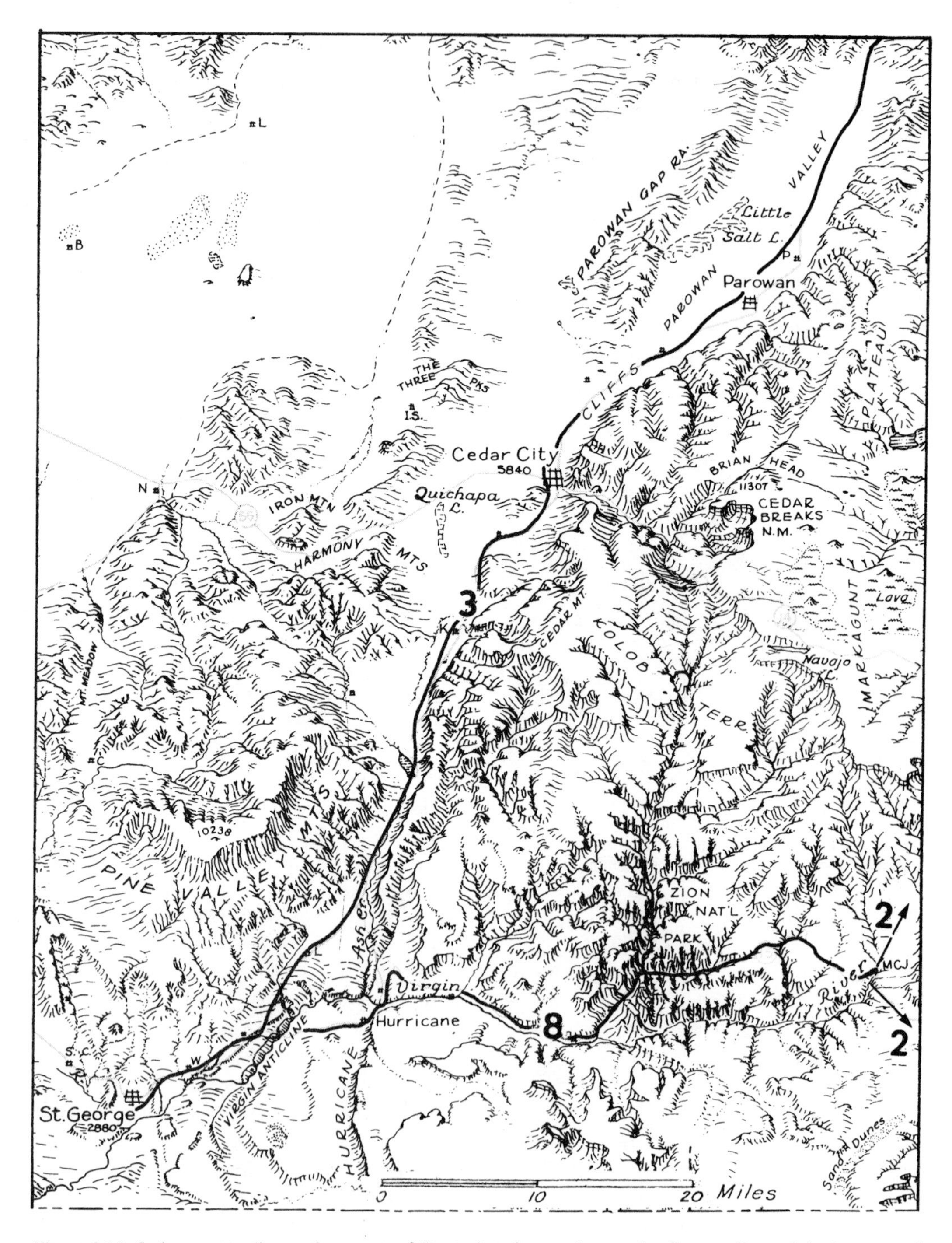

Figure 3.14. Index map to the southern part of Route 3 to its terminus at St. George. Route 8 leads eastward into the Zion National Park area and connects with Route 2 at Mt. Carmel Junction along U.S. Highway 89.

the fault escarpment where tan Cretaceous rocks are exposed beneath the pink Wasatch or Cedar Breaks Formation. The escarpment here is part of the Hurricane Cliffs that are along the Hurricane Fault and which are traceable as a topographic break from here well into Arizona.

182.9 Paragonah Exit. Cretaceous Kaiparowits rocks are exposed along the cliffs beneath the pink Wasatch Formation (fig. 3.15). Fault-repeated Wasatch beds are dropped down to near valley level at the base of the escarpment. Young basalt from a volcano on the plateau flank has covered the fault slices south of town.

Figure 3.15. View eastward of faulted upper Cretaceous and Wasatch beds near Paragonah. The Hurricane Fault trace is along the base of the escarpment.

187.8 Bridge Over Exit to Parowan. State Highway 143 leads southeast to Parowan and to Cedar Breaks National Monument. Pinkish Tertiary Wasatch Formation, purplish Tertiary volcanic rocks, and tan Cretaceous Kaiparowits Formation (fig. 3.16) are in fault-repeated exposures east of town. Volcanic rocks cap the Markagunt Plateau to the east. Cedar Breaks Monument is visible on the skyline to the southeast up Parowan Canyon.

190.6 Bridge over freeway at the southern edge of Parowan. Volcanic rubble and fault slices of Tertiary and Cretaceous rocks still form the west face of the Markagunt Plateau. West of the highway fault-repeated and down-faulted Wasatch or Claron Formation shows in Parowan Gap beyond Little Salt Lake playa.

197.2 Summit exit. South of here Tertiary volcanic rocks rest upon Cretaceous Kaiparowits Formation in fault slices that repeat the section, in general, downdropped toward the west.

199.8 Crest of hill north of Cedar City. Southwest and west from here are the multipeaked laccoliths of the Iron Springs mining district. Tertiary intrusions have domed into Mesozoic rocks and have reacted with the Jurassic Carmel Limestone to produce moderately extensive iron deposits of magnetite and hematite in the intrusion and around the intrusion borders. Three Peaks laccolith is to the west. Granite Mountain is beyond and in front of Iron Mountain which is to the southwest.

202.9 Double road cut through Tertiary volcanic rocks on a large spur leading out from the Hurricane Cliffs. Kaiparowits Formation is exposed at the escarpment.

205.2 Bridge over Union Pacific Railroad at the west end of **Cedar City.** Cedar Breaks National Monument in Tertiary Wasatch or Cedar Breaks Formation is visible on the skyline beyond the varicolored Jurassic and Triassic rocks which are exposed in Cedar Canyon. Mesozoic rocks are involved in pre-Hurricane Fault folds.

207.4 Bridge at the West Cedar City Interchange. College of Southern Utah is in the western edge of town east of the highway. Cross Hollow Hills to the southwest are composed of Tertiary and Quaternary volcanic rocks. To the southeast Cedar Mountin along the Hurricane Cliffs is capped by Cretaceous Straight Cliffs Sandstone. Trop-

SOUTHWESTERN UTAH
CEDAR CITY - ZION PARK - ST. GEORGE - IRON MINES

Period	Formation	Member	Thickness	Remarks
TRIASSIC	Kayenta Fm		700–1200	red mudstone; Shurtz Ss tongue; red mudstone
	Moenave Fm	Springdale Ss	110	pale red ss ledge
		Whitmore Pt M	15-70	gray slope
		Dinosaur Canyon M	400	light brown
	Chinle Fm	Pet Forest M	250-360	*petrified wood*
		Shinarump M	40-150	
	Moenkopi Formation (1810)	upper red	510	
		Shnabkaib Member	320	gypsum
		middle red	300	
		Virgin Ls M	130	*Tirolites*
		lower red	450	
		Timpoweap M	100	*Meekoceras*; UNCONFORMITY
PERMIAN	Kaibab Ls		850	*Dictyoclostus*
	Toroweap Fm		230	*Pugnoides*
	Conconino-Queantoweap Ss undivided		1250	
	Pakoon Fm		300	*Schwagerina*
PENN	Callville Ls		900	*Triticites*, *Fusilina*, *Fusilinella*, *Millerella*
MISS	Redwall Ls		960	*corals*, *brachiopods*; cherty
DEV	Crystal Pass ? Guilmette ? Simonson ? Dolomite		800	
	Ord-Sil? dolomite		200	cherty
CAMBRIAN	Muav Ls		1100	*unfossiliferous*
	Pioche Shale		215	*trilobite fragments*
	Prospect Mtn Quartzite		530	
P€	Vishnu Schist		—	1650 M.Y. K-Ar

Period	Formation	Member	Thickness	Remarks
Q	Basalt flows		0-500	6.0 M.Y. K-Ar Fortification basalt in Nevada caps Muddy Creek Fm
PLIO	Muddy Creek Fm		0-1400	
M	Page Ranch Volc		0-400	Three Peaks intrusive 21.1 M.Y.
OLIGOCENE	Rencher Fm		0-1000	21.5 M.Y.; tuff breccia, welded tuff, associated sediments
	Quichapa Fm		0-1600	22.0 M.Y.; ignimbrites; 24.0 M.Y.; 24.3 M.Y.
	Isom Ignimbrite		0-150	25.7 M.Y.
	Needles Range Tuffs		0-200	29 M.Y. K-Ar
PALEOCENE (Lower Eoc)	Cedar Breaks Fm (* max. thickness at Cedar Breaks; = Claron Fm of west Utah)		400–*1400	"White Claron"; "Red Claron"
CRETACEOUS	Iron Springs Formation	Kaiparowitz equiv.	4000±	
		Wahweap - Straight Cliffs equiv.		
		Tropic Shale equiv.		coal
		Dakota Ss equiv.		
JURASSIC	Carmel Fm	Winsor Mbr	180-320	
		Paria Riv Gyp M	70-150	"Curtis gypsum"
		Crystal Creek M	170	"Entrada"; *Pentacrinus*
		Kolob Ls Mbr	220-480	*clams*, *oysters*
	Temple Cap Mbr			
	Navajo Ss		2000–2300	sand dune cross-bedding

Figure 3.16. Stratigraphic section of rocks exposed in southwestern Utah in the Cedar City and St. George area (from Hintze, 1973).

ic Shale and Dakota Sandstone form the wooded slope below. Reddish slopes and ledges in the lower part of the escarpment are on the Carmel Formation. Navajo Sandstone is exposed as the white cliff-former above the reddish Triassic rocks near the base of the cliffs. Volcanoes high on the plateau provided basalt which has poured over the escarpment to the southeast of town (fig. 3.17).

Figure 3.17. Hummocky topography on dark lava flows which have poured over the Hurricane escarpment south of Cedar City at approximately Mile 208 from volcanoes high on the plateau.

211.9 Double road cuts through basalt which is offset by small faults with displacements of 10 to 15 feet. Crude columns are developed in the lava which has flowed over alluvial fan material obviously derived from the Triassic and Jurassic rocks to the east. Iron Mountain and its associated iron mines are visible to the west.

216.5 South of the Kanarraville Interchange, Jurassic Navajo Sandstone forms cliffs above folded Triassic rocks. Tertiary Wasatch Formation overlies Navajo Sandstone unconformably and in turn is overlain by Quaternary basalt.

219.9 Permian Kaibab Limestone appears at the base of the Hurricane Cliffs east of the Hurricane Fault trace (fig. 3.18). Reddish Triassic Moenkopi, Chinle and Moenave Formations form the slope below cliff-forming white Navajo Sandstone. Cretaceous Straight Cliffs Formation forms the caprock above the slope on the Dakota and Tropic Formations. Kaibab Limestone is overturned on the Kanarra Anticline.

Figure 3.18. View southeastward from south of Cedar City at approximately Mile 220 of Kaibab Limestone and older rocks exposed in the Hurricane Fault escarpment in the low hills, and the Navajo Sandstone and younger rocks in the White Cliffs escarpment along the skyline.

221.3 Rest area on the southbound lane. Kaibab Limestone is exposed near the base of the cliffs which are still capped by Cretaceous rocks. The Pine Valley Mountains to the southwest are carved in a large laccolith.

223.3 Fault slivers of red Triassic Moenkopi rocks are dropped down against Kaibab Limestone near the New Harmony exit.

223.9 Iron County—Washington County Line. Triassic Moenkopi Formation forms the reddish cliffs on the skyline to the east.

225.9 Bridge of the Kolob Canyon Inter-

change. Fins and buttresses of Navajo Sandstone show well on the skyline to the east up some of the canyons. Gray and tan Kaibab Limestone forms the lower set of cliffs associated with some fault slivers of red Moenkopi Formation.

229.0 Bridge over Ash Creek. The reservoir here does not hold water because of the jointed bedrock over which it was constructed. Black Ridge to the south is composed of Quaternary basalt which is the same age as basalt on the Hurricane Cliffs to the east, indicating the relatively recent major movement on the Hurricane Fault. Younger basalts have poured over the escarpment, however, indicating several periods of volcanic activity.

232.9 Underpass beneath **Snowfield Exit.** Northeast of the exit tan massive Coconino Sandstone is visible below gray Toroweap Formation and tan Kaibab Limestone in the cliffs. Pink and red beds at the base are Triassic rocks. The Kanarra fold shows in various attitudes of the Kaibab Limestone. Most of the Interstate highway is still built over young basalt or on debris swept out from the Pine Valley Mountains to the west (fig. 3.19).

234.5 **Bridge at the Pintura Exit.** Coarse bouldery debris on the west side of the road is mudflow or alluvial fan material from the Pine Valley Mountains. East of the road westerly dipping Kaibab Limestone is exposed next to the trace of the Hurricane Fault.

237.9 Steeply folded Kaibab Limestone (fig. 3.20) is exposed on the east side of the Kanarra fold and the Hurricane Fault.

239.0 **Cross Over Toquerville—Zion National Park Exit to Utah State Highway 17.** Quaternary basalt caps Navajo Sandstone on both sides of the road as part of inverted valleys.

241.9 Navajo Sandstone and younger rocks are exposed to the west in front of the Pine Valley Mountain intrusion. Shinarump-capped terrace above the Moenkopi Forma-

Figure 3.19. The northern part of the Pine Valley Mountains as seen westward from near the Snowfield exit at approximately Mile 231. The Pine Valley Mountains are carved out of a large igneous intrusion which appears to be laccolithic. Tree-covered slopes in the foreground are alluvial fans which are spreading eastward from the Pine Valley Mountains.

Figure 3.20. Steep eastward-dipping Kaibab Limestone forming the prominent cliff at the top of the Hurricane escarpment. Toroweap and Coconino Formations are exposed in the less well-bedded lower parts of the faulted face. Steep dips are eastward off the Kanarra Fold which diagonals across the fault escarpment. The Hurricane Fault trace is at the base of light-colored exposures near the brush line.

tion rises above the Kaibab exposures in the Hurricane Cliffs to the east.

242.9 Bridge over access road north of Leeds. The small mine dump to the east is in the Silver Reef Sandstone Member of the Chinle Formation. Toward the west the Moenave Formation forms the brick red sandstone cliffs in front of the higher and lighter colored Navajo Sandstone, near the base of the Pine Valley Mountains.

243.5 Bend in the highway. Silver Reef Sandstone caps the cuesta to the northwest and the middle of three prominent sandstones in the Chinle Formation crosses the road approximately here and forms the cuesta to the southwest. Shinarump Sandstone member is exposed on to the southeast. Chinle beds begin to swing from the west flank over the northern nose of the Harrisburg Dome to the east. Beyond the bend a short distance, dumps of some of the mines of the Silver Reef district (fig. 3.21) are visible to the southwest in the sandstone cuesta.

244.1 Leeds Interchange. Leeds was an old silver mining district where fossil leaves and other organic material were replaced by native silver or copper-uranium-vanadium minerals. The ore was mined from the Silver Reef Sandstone Member of the Chinle Formation which occurs approximately 175 feet above the Shinarump Sandstone Member. To 1953 production from the district approached eight million dollars. To the south of the interchange the freeway continues in a subsequent valley between cuestas formed of the two lower sandstone members of the Chinle Formation.

245.5 A gap in the cuesta along a small drainage shows the mining area of the Silver Reef District to the northwest.

246.3 Double road cuts on both lanes through varicolored lower mudstones of the Chinle Formation above the Shinarump Conglomerate which holds up the cuesta to the east. Interstate Highway 15 continues toward the southwest across alluvial fill and poor exposures.

Figure 3.21. View southeastward into the northern part of the Silver Reef Mining District from approximately Mile 243. Rocks in the foreground and near the mine dump are in the Chinle Formation. The ledge-forming sandstone above the dump is the Silver Reef Sandstone Member of the Chinle Formation and was the main ore bearer for the silver, copper, uranium, and vanadium minerals in the district.

247.7 View toward the north and northwest from near the brow of the hill shows the Silver Reef Mine dumps in the cuesta a short distance west of the highway.

249.2 Double road cuts through coarse terrace gravel deposited across a cut surface on Moenave Formation. The gravel is largely of igneous pebbles and boulders derived from the Pine Valley Mountains to the northwest. The prominent cuesta to the east is held up by the Shinarump Member. Contact of the Chinle and Moenave Formations is somewhere in the middle of the alluvial-blanketed valley at approximately the position of the highway. The highway swings onto Moenave beds toward the southwest.

250.6 Bridge Over Interchange of Utah State Highway 15 at Hurricane and Zion National Park Turnoff. For a description of the geology along Utah State Highway 15 east to Zion Canyon and Mt. Carmel Junction see

Geologic Guide Segment 8. The prominent red sandstone toward the northwest is the Springdale Member of the Moenave Formation and the interchange is on the less well-exposed Dinosaur Canyon Member of the formation. Light-colored Navajo Sandstone forms the next series of cliffs to the northwest around the base of the Pine Valley Mountains (fig. 3.22).

Figure 3.22. View northwestward from approximately Mile 250 of the Pine Valley Mountains and their foothill belt. Relatively flat country in the foreground is a gravel cap over a pediment carved on Chinle Formation. Navajo Sandstone forms the prominent light-colored exposures at the top of the cuesta in the middle distance. The Pine Valley intrusion is exposed in the high peak of the mountains on the skyline.

252.0 Deep double road cuts through the middle part of the Moenave Formation. Sandstone near the top is lower Springdale Member. The basalt which caps the ridge shows excellent columnar jointing and rests upon a lag gravel over the Moenave beds (fig. 3.23). The linear basalt ridge represents an inverted valley. When the basalt flowed into the Virgin River Valley from the northwest it flowed down a valley and protected it from subsequent major erosion. Softer Moenave rocks on either side of the lava flow have been removed by subsequent erosion, by the displaced stream and associated streams, leaving the former valley now standing high.

Figure 3.23. Colunnar jointed basalt flow which delimits an inverted valley. Basalt initially flowed down a low point in the topography but because of its resistance to erosion it has protected the valley while more easily eroded beds on either side have been removed. Thus the former valley now rises as a ridge. Basalt here rests on red Moenave beds.

253.6 Cross drainage and bridge over access road in eastern Washington. Spring area a short distance to the west is along the Washington Fault which has dropped upper Moenave beds on the west down against lower Moenave beds on the east.

254.7 Cross Bridge In Washington—Middleton Interchange. Springdale Sandstone forms the prominent reddish bluff to the north and continues north of the highway westward to St. George.

256.5 Deep stepped road cuts through Moenave Formation (fig. 3.24) capped by another basalt-armoured inverted valley. This basalt is less well jointed than that to the east but the linear ridge apparently formed in the same way. Moenave beds here appear similar to rocks interpreted elsewhere to have been deposited on tidal flats of an arid coastline.

Figure 3.24. Crudely jointed basalt capping an extensive exposure of the lower part of the Moenave Formation at Mile 256.5. The long basalt ridge expresses an inverted valley, where basalt protected the gravelly sediments deposited in the valley fill while unprotected areas along either side of the valley bottom have now been eroded away to produce the lowlands on either side of the long narrow ridge. The flow extends southeastward from a volcanic field along the flanks of the Pine Valley Mountains. Well-bedded Moenave rocks are thought to have been deposited on a tidal flat during the late Triassic.

257.1 Center of Bridge Over Interstate 15 at the Interchange East of St. George. The access road which separates from the freeway short of the bridge leads into the eastern end of the main business street in St. George, noted for its white Mormon Temple (fig. 3.25).

Figure 3.25. Mormon Temple in St. George. The temple was built of red Moenave Sandstone but has been painted white.

Segment 4

0.0 North Cove Fort Interchange. Leave the freeway and head east toward the south end of the Pavant Range and U.S. Highway 89 (fig. 4.1). To the southwest, the high conical hill is a cinder cone related to some of the recent basaltic volcanism of the area. East of the freeway interchange the road is built across volcanic debris.

3.4 Cove Fort. Cove Fort (fig. 4.2) was built by Brigham Young in 1867 as a stopover point between Salt Lake City and St. George. It is constructed of basalt quarried in the low hills to the west, and has pinkish andesitic rocks for chimneys for the numerous fireplaces of the fort.

4.4 Pass beneath Interstate Highway 70 interchange and head east on the entrance road to join the highway. Sulfurdale, a short distance to the south of the interchange, was a major source of sulfur in early pioneer days and rocks in the open pits still show considerable elemental sulfur which fills voids in some of the volcanic rocks. The deposit appears to be related to recently active fumaroles.

5.9 Boundary of Fish Lake National Forest in pinion and juniper woods. The canyon along which the route travels separates the Tushar Range on the south from the Pavant Range on the north. Bleached and altered volcanic rocks are exposed in low road cuts in both the east- and west-bound lanes.

9.3 Ashy-appearing tuffaceous rocks exposed in cuts on the eastbound lane. The general southeastward dip of the rocks is evident in the cuts and suggests that the Tushar Range is composed of thousands of feet of volcanic rocks.

10.5 Summit and divide between Cove Creek on the west and Clear Creek on the east. Road cuts show ashy volcanic rocks characteristic of the Tushar Range and the Marysvale district. The view to the east is principally of the northern flank of Mount Belknap and the volcanic pile of the Tushar Range.

13.2 Extensive coarse Sevier River deposits are well exposed in road cuts on the north (fig. 4.3). These exposures are typical of bedrock between the summit and here and show the very coarse nature of the accumulation.

14.2 Excellent exposures of tuffaceous mudflow or lahar deposits occur in ledges to the north of the road.

17.1 Narrows of Clear Creek. Vertical walls are cut in ignimbrites which here show only weak columnar jointing. The canyon, at present (1976) is just wide enough for a two-lane road and the creek.

18.0 Welded ignimbrite of the lower canyon walls shows strong columnar jointing. Such ignimbrites are common in the lower part of the volcanic section to be seen along Clear Creek Canyon between here and

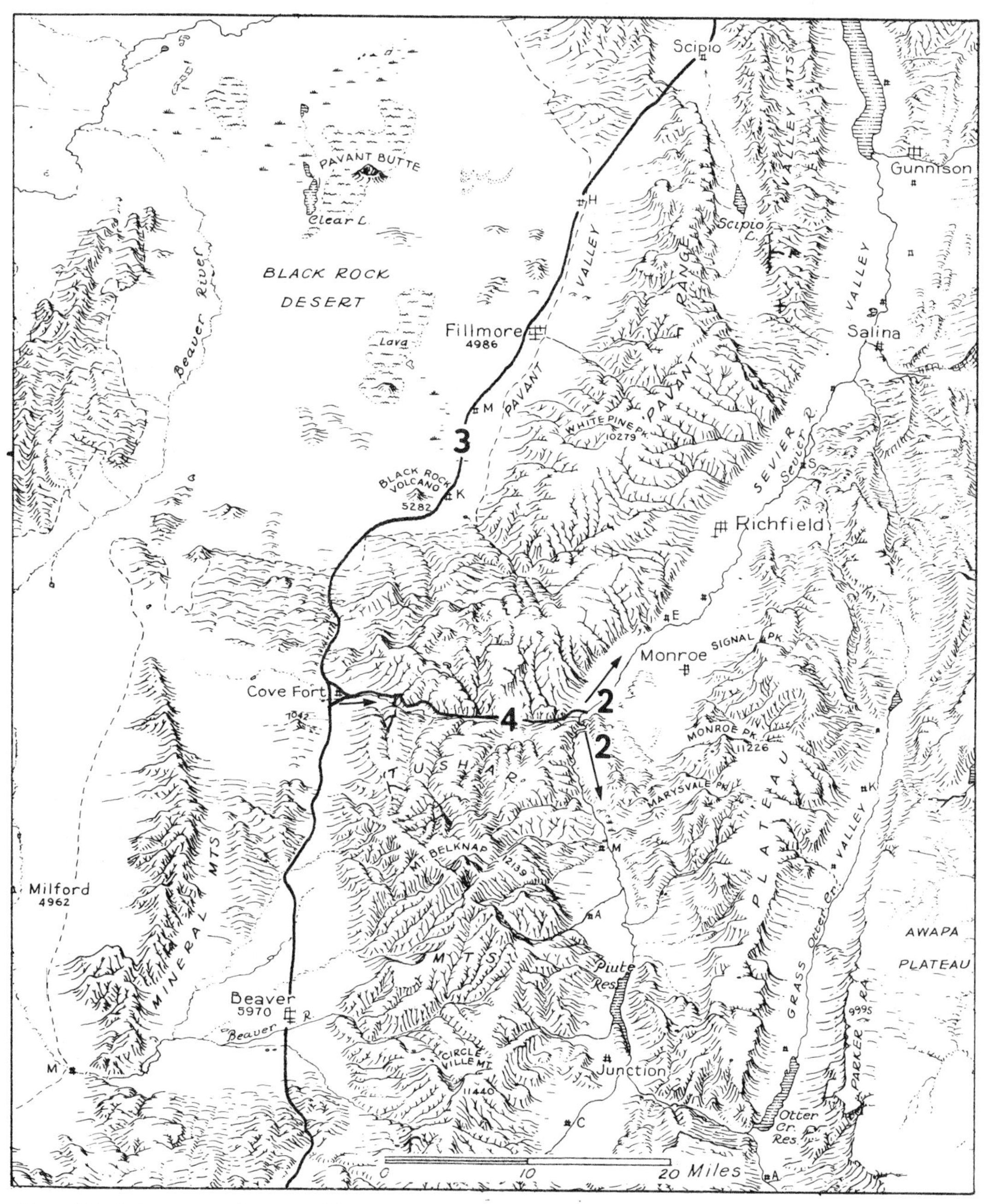

Figure 4.1. Index map to Route 4 which connects westward to Route 3 and eastward to Route 2. Route 4 is along Interstate Highway 70 which has been constructed through a low pass that separates the Pavant Range, on the north, from the Tushar Mountains, on the south.

U.S. Highway 89, 5 miles to the east. Side canyons to the south and north provide glimpses to the section and suggest that the rocks near the road are only part of a sequence of rocks 2,000 to 3,000 feet which are exposed here.

22.0 Cliff in spectacular exposures of jointed ignimbrite (fig. 4.4). The exposure may be formed in three flow units. Sharp needles are produced by weathering of the same rocks on the south side of the canyon opposite the ranch house.

Figure 4.2. Cove Fort which was built for Brigham Young as a stop-over point on the road between Salt Lake City and St. George. The fort is built, in large part, of basalt which was quarried from flows to the west and has fireplaces of pinkish andesite which was quarried from the mountains to the east.

Figure 4.4. Prominently jointed cliffs of welded ignimbrite, as seen eastward from approximately Mile 21. Two or three flow units are exposed here in the Oligocene volcanic section.

Figure 4.3. Coarse Sevier River Formation exposed in roadcuts at Mile 13.2, on the east side of the summit in the headwaters of Clear Creek. The Sevier River deposits in general rest unconformably on some of the older volcanic and clastic units in the mountains.

Figure 4.5. Various volcanic units exposed near the mouth of Clear Creek Canyon, as seen westward from near the connection of Utah State Highway 13—Interstate 70 with U.S. Highway 89. Ashy units generally weather to more rounded slopes and welded ignimbrites form the more resistant ledges.

24.5 Junction of Utah State Highway 13 and Interstate 70 with U.S. Highway 89. For a description of the routes north and south from here see Geologic Guide Segment 2. Joseph and Elsinore are to the north and Marysvale and Junction are to the south. Strong east dip of the volcanic sequence shows in the cuestas to the west (fig. 4.5) and north.

Segment 5

Guide along Interstate 70 from Green River to Salina and U.S. Highway 89

0.0 West Edge of Green River. Interstate Highway 70 is constructed across Mancos Shale and valley fill along Saleratus Creek.

Erosion of gently dipping Ferron Sandstone (fig. 5.1) produces rounded cuesta to the south which rises above the general erosional level on the Mancos Shale. Fossil ammonoids and various molluscs can be collected out of the laminated beds immediately below the silty remnant of Ferron Sandstone all along the outcrop band.

2.4 Cross Beneath Bridge of U.S. Highway 50-6 Interchange. U.S. Highway 50-6 leads northwestward toward Price, Provo, and Salt Lake City. For a guide to this section consult Guide Segment 1. Guide Segment 5 proceeds westward across the San Rafael Swell, visible ahead, and through the Wasatch Plateau to Salina. Beyond the interchange Interstate 70 climbs the gentle cuestas produced in Cretaceous beds along the east margin of the San Rafael Swell, a large domal structure rimmed by hogbacks and cuestas in Permian to Cretaceous rocks in east central Utah.

Flatirons and serrate cockscombs on the east flank of the San Rafael Swell are visible ahead in reddish Wingate Sandstone and white Navajo Sandstone. San Rafael River crosses through the "Reef" through the notch at approximately one-thirty ahead.

7.7 Crest of Ferron Sandstone cuesta. View to north and behind is of Book Cliffs and East and West Tavaputs Plateaus (fig. 5.2), separated by Desolation Canyon, the gorge of the Green River.

In the double road cuts on both the east- and westbound lanes white volcanic ash beds have been faulted and also mark the base of the thin silty Ferron Sandstone and the top of a fossiliferous laminated part of the lower Mancos Shale. Ammonoids and molluscs of various types can be collected from the platy shaly beds (fig. 5.3). The fossils are very fragile, however, and do not transport well. These fossiliferous shale and siltstone beds and the overlying thin silty sandstone are the distal edge of a thick sandstone and coal-bearing deltaic sequence which is well exposed along I-70 on the west flank of the San Rafael Swell.

8.7 Oyster beds composed of *Gryphaea newberryi* (fig. 5.4) form low rounded exposures in road cuts at the base of the Mancos Shale. In other areas equivalent oyster beds have been utilized for gravel because of the great abundance of the fossils.

Mancos Shale here rests upon gray green,

GREEN RIVER AREA, UTAH

Subsurface thicknesses from well data.

System	Series	Unit	Member/Formation	Thickness (ft)	Notes
JURASSIC		San Rafael Group	Entrada Ss	520	Permo-Jurassic wind directions- Poole (1962)
JURASSIC		San Rafael Group	Carmel Fm	120	
JURASSIC		Glen Canyon Group	Navajo Ss	450	
TRIASSIC		Glen Canyon Group	Kayenta Fm	170-270	
TRIASSIC		Glen Canyon Group	Wingate Ss	300–400	
TRIASSIC		Chinle Fm		240-300	Chinle regional: Stewart (1969)
TRIASSIC		Chinle Fm	Moss Back Cg M	40-140	
TRIASSIC		Moenkopi Fm	upper memb	500	
TRIASSIC		Moenkopi Fm	Sinbad Ls M	50	
TRIASSIC		Moenkopi Fm	lower memb	150	
PERMIAN	Leon	Kaibab Limestone		60-160	
PERMIAN	Leon	Cutler Group	White Rim Ss	300	Coconino of some well logs
PERMIAN	Wolfcampian	Cutler Group	Organ Rk Sh	0-200	
PERMIAN	Wolfcampian	Cutler Group	Cedar Mesa Ss	200–400	
PERMIAN	Wolfcampian	Elephant Canyon Formation		600	*Pseudoschwagerina Dunbarinella Triticites Schwagerina* "Rico Fm" of older reports
PENNSYLVANIAN	Missourian	Hermosa Group	Honaker Trail Fm	800–1200	
PENNSYLVANIAN	Desmoinsian	Hermosa Group	Paradox FM	0–3000	salt
PENNSYLVANIAN	M-A	Molas Fm		50	
MISS		Redwall Ls (Leadville of Colo.)		500–700	*endothyrid foraminifera, corals, crinoid columnals*
DEV		Ouray Ls		100	
DEV		Elbert Fm		230	
DEV		Elbert Fm	McCracken Ss M	0-40	UNCONFORMITY
CAMBRIAN		Cambrian limestones undifferentiated		800–1000	"Lynch" Ls; "Maxfield" Ls
CAMBRIAN		Ophir Fm		200	
CAMBRIAN		Tintic Qtzt		160-200	
P-Є		Precambrian granite		—	1.8 B.Y. Rb-Sr

System	Group	Formation	Member	Thickness (ft)	Notes
EOCENE		Green River Fm		0–7000	oil shale, fresh water ls and shale
PALEOCENE	"WASATCH GROUP"	Colton Fm ("Wasatch" of some authors)		1500	reddish continental mudstone, siltstone, and sandstone
PALEOCENE	"WASATCH GROUP"	Flagstaff Ls		0-100	fresh water ls
CRETACEOUS		North Horn Fm (Tuscher Fm)		400– 500	UNCONFORMITY
CRETACEOUS	"MESAVERDE GROUP"	Price River Formation	Farrer facies (generally noncoal-bearing)	725	
CRETACEOUS	"MESAVERDE GROUP"	Price River Formation	Neslen facies (coal bearing)	200– 250	1' coal
CRETACEOUS	"MESAVERDE GROUP"	Price River Formation	Sego Ss M	50-150	
CRETACEOUS	"MESAVERDE GROUP"	Price River Formation	Buck Sh Tong	0-50	Mancos tongue
CRETACEOUS	"MESAVERDE GROUP"	Blackhawk	Castlegate Ss M	85	two 3' coals
CRETACEOUS	"MESAVERDE GROUP"	Blackhawk	Deseret Memb	140	6' coal
CRETACEOUS	"MESAVERDE GROUP"	Blackhawk	Grassy Memb	200-250	pinches out to east
CRETACEOUS	"MESAVERDE GROUP"	Blackhawk	Mancos Sh Ton	100-150	
CRETACEOUS	"MESAVERDE GROUP"	Blackhawk	Sunnyside M	200-0	*Nucula*
CRETACEOUS		Mancos Shale	Pierre fauna; Telegraph Creek fauna; Niobrara fauna	3000	*Scaphites hippo*; *Inoceramus*; *Inoceramus lobatus*; *Baculites aquilaensis*; *Desmoscaphites bass*
CRETACEOUS		Mancos Shale	Ferron Ss M	10	*Inoceramus Ostrea Baculites Collignoniceras*
CRETACEOUS		Mancos Shale	Tununk Shale Member	800	*Ostrea (Carlile age)*
CRETACEOUS		Dakota Ss		0-40	*Gryphaea newberryi*
CRETACEOUS		Cedar Mtn Fm		250	Buckhorn Cg Mbr
JURASSIC		Morrison Fm	Brushy Basin	300	green and purple dinosaur beds
JURASSIC		Morrison Fm	Salt Wash M	180	
JURASSIC	San Rafael Group cont'd		Summerville	20-130	
JURASSIC	San Rafael Group cont'd		Curtis Fm	0-75	chocolate beds
JURASSIC	San Rafael Group cont'd		Entrada Ss	(520)	Arches Nat'l Mon

Figure 5.1. Stratigraphic section of formations in the Green River area in east central Utah (fron Hintze, 1973).

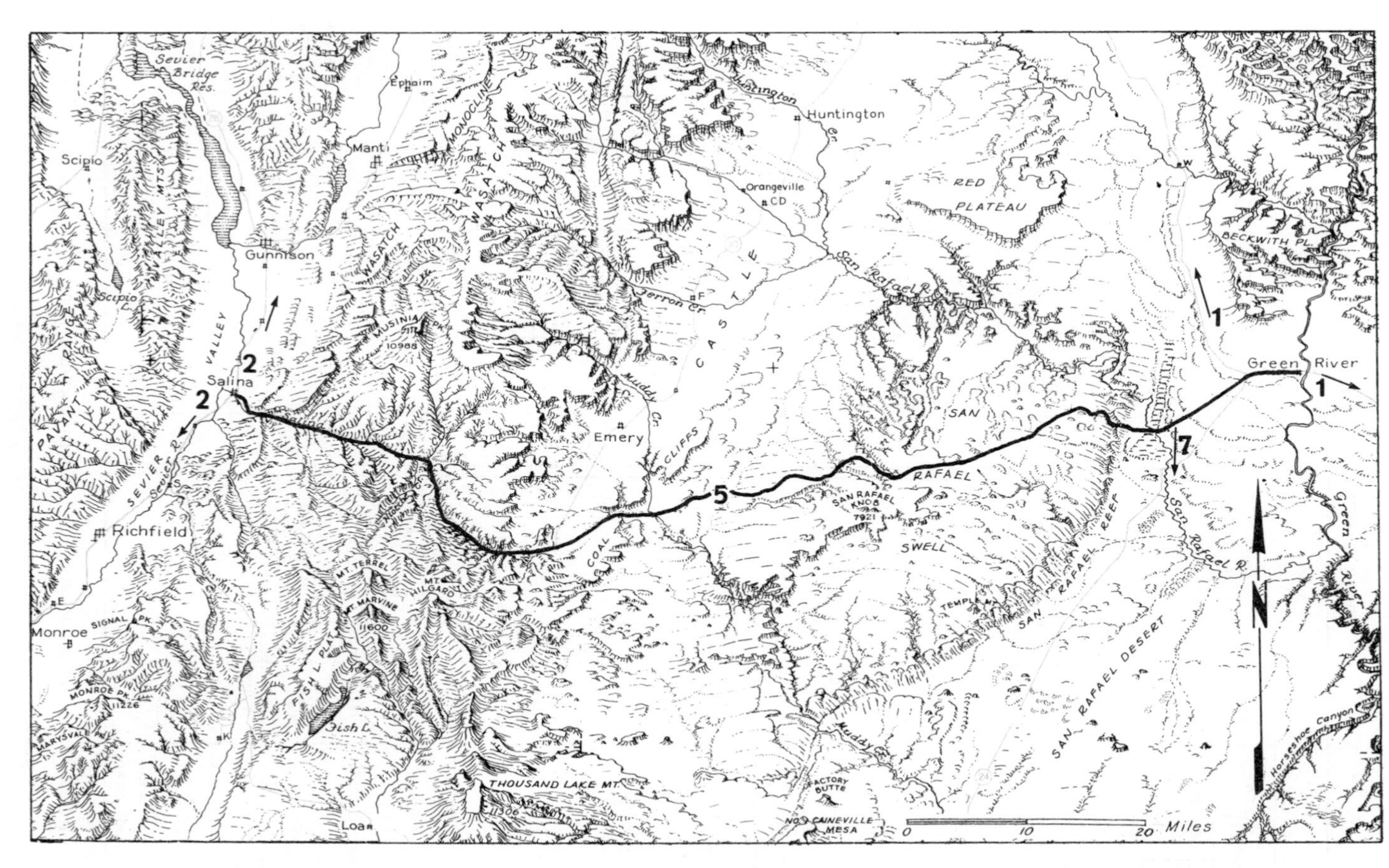

Figure 5.2. Index map of Route 5 between Green River and Salina, across the San Rafael Swell and the south end of the Wasatch Plateau.

Figure 5.3. Exposures of Mancos Shale in road cuts along Interstate Highway 70 at Mile 8. Thin-bedded units at the top are in the Ferron Sandstone. Laminated beds between the massive, light-colored lower exposures and the Ferron Sandstone contain abundant flattened ammonoids and bivalves.

Figure 5.4. Weathered exposures of the *Gryphaea newberryi* beds at Mile 18.7. These oyster beds are the basal unit of the Mancos Shale and rest on lacustrine and fluvial lower Cretaceous Cedar Mountain Formation.

ashy-appearing Cedar Mountain Formation of Lower Cretaceous age (fig. 5.1). The resistant conglomeratic basal Buckhorn Member of the Cedar Mountain Formation is exposed in cuts a short distance west along the highway and separates the more greenish Cedar Mountain beds above from the maroon and purplish gray beds of the Morrison Formation below.

10.0 Bridge of the Hanksville Interchange (fig. 5.5.). Utah State Highway 24 leads south to Hanksville and the upper reaches of Lake Powell. For a guide to this state highway see Guide Segment 7, which also covers the road through Capitol Reef National Monument and westward to join U.S. Highway 89 near Sigurd.

Morrison Formation is exposed both east and west of the interchange. Lake and deltaic deposits of the Brushy Basin Member (fig. 5.1) form purplish rounded exposures to the east. Ancient stream channels which are filled with light gray sandstone weather out into three-dimensional relief in the Salt

Figure 5.5. View eastward from the Hanksville Interchange along State Highway 70 at Mile 10 toward Cretaceous exposures in the distance. Linear ridges in the foreground are held up by sandstone fillings of channels in the Morrison Formation. The cuesta in the middle distance is on the Ferron Sandstone in the lower part of the Mancos Shale (photograph by W. K. Hamblin).

Wash Member (fig. 5.1) west of the interchange. Softer floodplain deposits have been eroded away from some channel fills to leave them as meandering sinuous ridges that mark the old stream courses.

11.3 Bridge Across the San Rafael River. Summerville "castle walls" are visible both south and north of the bridge, capped by a massive thick bed of gypsum (fig. 5.6). Thick lenticular stream-deposited sandstone beds of the lower Morrison Formation cap the ridges. The Summerville Formation was deposited as part of an extensive arid-coast tidal flat. Thin gypsum stringers through the formation and the massive gypsum at the top were deposited by evaporation of seawater along the margin of a sea which extended from near here northward along the Rocky Mountain area into northern Canada. West of the bridge greenish Curtis Sandstone, deposited in a late marine tongue of the Sundance Sea, forms low sandy exposures south of the road.

Figure 5.6. View southward from bridges over the San Rafael River along Interstate Highway 70 to the Summerville Formation (S) and the overlying Morrison Formation (M). Light-colored rocks beneath the Summerville Formation at the extreme right are upper beds of the Curtis Sandstone. A thick gypsum bed at the top of the Summerville Formation caps the castellate cliffs.

12.8 Rest areas here are built on Carmel Formation, the unit which forms the resistant flatirons and resistant ridges (fig. 5.7) at the base of the San Rafael Reef north and south of the highway (fig. 5.8). Lower limestones of the unit are fossiliferous but upper beds are not and record gradual increase in salinity above that in normal seas until gypsum was precipitated in what were evaporating pans along the shoreline.

Figure 5.7. View northward along the monocline on the east side of the San Rafael Swell from the rest areas at approximately Mile 12.8. Dark flatirons are held by lower fossiliferous limestones of the Carmel Formation which here rest on light-colored, massive Navajo Sandstone.

The overlying Entrada Formation is relatively soft here and has helped to produce a prominent strike valley parallel to the Reef (fig. 5.8). Entrada beds are exposed in cliffs south of the rest area where they form castellate features like the Summerville Formation above. They are separated in the cliff by the greenish Curtis Formation. Entrada rocks here are thought to be tidal flat deposits, like those of the Summerville Formation.

West of the rest areas, Interstate 70 has been blasted through deep narrow cuts in the

Figure 5.8. View southward along The Reef on the east flank of the San Rafael Swell from near the rest areas at Mile 12.8. Navajo Sandstone forms the top of the cockscomb along the monocline. Permian and Triassic rocks are exposed in the center of the uplift, toward the right, and are nearly flat lying away from the steep monoclinal flexure. Light-colored Morrison Formation caps the tableland on the left of the photograph, above the dark ledges on the Summerville Formation. Light-colored Curtis Sandstone separates the Summerville Formation from the underlying Entrada and the banded Carmel Formations which occur in the lowlands between the monocline and the tablelands (photograph by W.K. Hamblin).

underlying Navajo, Kayenta, and Wingate Sandstone which stand high on the monoclinal fold at the east of the San Rafael Swell (fig. 5.9). Wingate and Navajo Sandstones are both cross-bedded windblown sandstones. They are separated by the more flaggy-bedded, stream-worked Kayenta Sandstone which forms a shoulder and brushy area midway through the narrow cut.

13.4 West end of the narrow cuts. Top of the stream-deposited Chinle Formation and base of the Wingate Sandstone. Chinle beds are maroon and reddish brown shale and siltstone and also include the massive sandstone ledge on top of the prominent slope-forming reddish mudstone (fig. 5.10). Thin bedded Moenkopi Formation is exposed below the mudstone to the west and lacks the thick sandstone beds of the Chinle. Chinle rocks have produced numerous fossil reptiles and considerable fossil plant material. In some areas the formation has also been a significant producer of uranium minerals, usually deposited as replacements of organic plant debris.

From here the highway climbs up the east

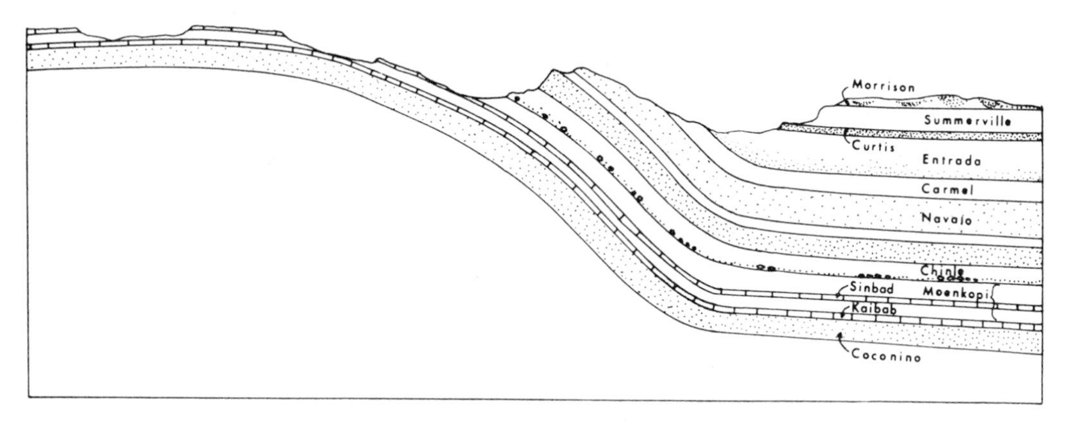

Figure 5.9 Generalized east-west geologic cross section across the monoclinal fold of the east edge of the San Rafael Swell approximately along Interstate 70. The cross section is drawn as though looking toward the north.

SAN RAFAEL SWELL

System	Group	Formation	Thickness	Remarks
PERMIAN		Kaibab Ls	0-85	Park City equivalent
PERMIAN		"Coconino" Ss = undivided White Rim and Cedar Mesa Ss of Cutler	600—800	Diamond Creek Ss of Wasatch Mtns
PERMIAN		Elephant Canyon Fm	500—700	pre-Permian known from limited well data
PENNSYLVANIAN	Hermosa Group	Honaker Trail Fm	200—500	
PENNSYLVANIAN	Hermosa Group	Paradox Fm?	300—500	Oquirrh Fm equivalents
PENNSYLVANIAN	Hermosa Group	Pinkerton Trail Fm	100—200	Manning Cyn equiv
MISSISSIPPIAN		Redwall Ls	600—900	Deseret Ls equivalent; Gardison Ls equivalent
DEV		Ouray Ls	100—200	
DEV		Elbert Fm	200—400	
CAMBRIAN		"Lynch" Dolomite; "Maxfield" Limestone	1000—1300	
CAMBRIAN		Ophir Fm	200	
CAMBRIAN		Tintic Quartzite	150—300	
P€		"granite"	—	

System	Formation	Member	Thickness	Remarks
CRETACEOUS	Mancos Shale	Blue Gate Shale M	1500+	
CRETACEOUS	Mancos Shale	Ferron Ss Member	100—300	12' coal near Ferron
CRETACEOUS	Mancos Shale	Tununk Shale Member	400—700	
CRETACEOUS	Dakota Ss		0-50	
CRETACEOUS	Cedar Mtn Fm		0-100	*"gizzard stones"*
JURASSIC	Morrison Fm	Brushy Basin Sh M	250—400	*dinosaur bones*
JURASSIC	Morrison Fm	Salt Wash Ss M	200—300	uranium in Green River mining district
JURASSIC	Summerville Fm		260—330	chocolate siltstones gypsum veinlets
JURASSIC	Curtis Fm		75—250	*Pentacrinus* *Camptonectes*
JURASSIC	Entrada Ss		300—800	poorly cemented earth- source of modern dune sands of south San Raf- ael area
JURASSIC	Carmel Fm (=Twin Creek, lower Arapien)		150—650	gypsum; red and green shale; *Trigonia* *Ostrea* *Camptonectes*
JURASSIC – TRIASSIC	Navajo Ss		440—540	highly crossbedded
TRIASSIC	Kayenta Fm		40—240	these sandstones form the reef along the southeastern margin of the San Rafael swell
TRIASSIC	Wingate Ss		350-400	
TRIASSIC	Chinle Fm	upper members	130-280	red and purple silt- stone and fine ss
TRIASSIC	Chinle Fm	Moss Back M	0-170	
TRIASSIC	Chinle Fm	"Temple Mtn" M	0-100	red and green muds
TRIASSIC	Moenkopi Fm	upper member	430	reddish brown to tan siltstone and silty sandstone
TRIASSIC	Moenkopi Fm	Sinbad Ls M	50-150	*Meekoceras*
TRIASSIC	Moenkopi Fm	lower memb	140-200	green pyritic silts

Figure 5.10. Stratigraphic section of rocks exposed or encountered in wells in the San Rafael Swell Area (from Hintze, 1973).

flank of the San Rafael Swell through Moenkopi and older Kaibab beds (fig. 5.1).

15.9 Rest Areas for both east- and westbound lanes. Rest area for westbound lane is built on the upper Kaibab Limestone. White Rim or Coconino Sandstone (fig. 5.10) is visible in the gorge to the north. The rest area for the eastbound lane is built on the Sinbad Limestone of the Moenkopi Formation and looks eastward over the narrow cut through the San Rafael Reef to the gray slopes of Mancos Shale and the East Tavaputs Plateau in the distance (fig. 5.11).

Figure 5.11. View eastward from the rest area of the eastbound land showing the highway route through the narrows along the east side of the San Rafael Swell. Kaibab (K) Limestone is the oldest unit exposed in this area and is overlain by light-colored and dark-colored Moenkopi Formation (M) that forms the slope zone near the base of the excarpment. Chinle Formation (C) forms the upper banded exposures and slope zones beneath the prominent Wingate Sandstone cliffs that hold up the high peaks of the cockscomb along the monocline. Younger Jurassic rocks are exposed through the deep V-shaped notch and Cretaceous rocks form exposures in the far distance in the Book Cliffs area.

Angular cliffs of Wingate Sandstone above reddish brown Chinle and Moenkopi beds form much of the scenery along the eastern flank of the San Rafael Swell and contrast with the tan to light gray or gray green lowlands on the top of the Kaibab and in lower Moenkopi beds.

17.7 Long road cut on the east side of the eastbound lane shows tan Moekopi beds resting disconformably upon cherty lag-gravel filled channels cut into the underlying Kaibab Limestone (fig. 5.12). Light tan to gray chert obviously was derived from weathering of Kaibab beds. The westbound lane of I-70 is on the top of the Kaibab Limestone, at the unconformity.

Figure 5.12. Disconformable contact between the Kaibab Limestone (K) and the Moenkopi Formation (M) along the east side of the freeway at Mile 17.7. Channels filled with lag deposits of sharp chert fragments occur at the unconformity. Much of late Permian and early Triassic time is represented by erosion at the unconformable surface.

19.9 Crest of the steep climb up the flexed eastern monoclinal limb of the San Rafael Swell. The grade now flattens westward and has been constructed across lowermost Moenkopi beds. Sinbad Limestone forms the cuesta caps above the road. Kaibab Limestone is still exposed in canyons below the road to the south and southeast.

22.0 Basal beds of Sinbad Limestone are exposed in road cuts. Beyond the sweeping curve the road rises onto the stripped upper surface of the Sinbad Limestone and continues ahead on this gentle surface over the crest of the San Rafael Swell.

29.8 Prominent sandstone cliff to the north is held up by the Moss Back Sandstone Member of the Chinle Formation (fig. 5.13). Road in the upper reddish Moenkopi beds leads to Uranium "glory holes" dug in the lower part of the Moss Back Cliff. To the west the road climbs slowly onto upper Moenkopi redbeds and then cuts up through the complexly lenticular sandstone of the Moss Back Member. The sandstone is probably a deposit of a sediment-charged braided stream system. Beyond the cut, to the west, poor exposures of soft Chinle Formation can be seen in some road cuts.

32.1 Pass between rounded yellow brown outcrops of cross-bedded Wingate Sandstone. Wingate beds form the joint-controlled fins in the cliffs north of the road. West of here the road rises up through Wingate Sandstone exposures.

34.1 San Rafael Knobs Rest Area. The knobs are in lowermost Navajo Sandstone with the woodlands at the road on more slabby-bedded Kayenta Sandstone. Canyons northwest of the rest area expose Chinle Beds in their lower channels, vertical angular cliffs of Wingate Sandstone, and upper slabby-bedded Kayenta Sandstone at the canyon rim (fig. 5.14). Cliffs of rounded Navajo Sandstone of the western flank of the San Rafael Swell rim the juniper-covered Kayenta beds in the background.

Figure 5.13. Uppermost Moenkopi beds form the slope in the foreground beneath the cliff-forming Shinarump Sandstone Member near the base of the Chinle Formation at approximately Mile 30. Locally the Shinarump Sandstone has produced minor amounts of uranium minerals and its outcrops were extensively prospected during the uranium boom in the 1950s. Moss Back Sandstone is thought to be the deposits of braided stream systems deposited on top of tidal deposits of the Moenkopi Formation.

Figure 5.14 View northwestward from the San Rafael Knobs rest area toward the northwestern margin of the San Rafael Swell. Chinle Beds (C) are exposed in the lower slopes within the canyon beneath the massive Wingate Sandstone cliff (W). Kayenta Sandstone forms the juniper-covered tablelands between Wingate Sandstone and overlying white cliffs of Navajo Sandstone (N). The Wasatch Plateau rises in the far distance along the skyline.

36.6 Cross Bridge Over Eagle Canyon. Vertical walls of gorge are in Wingate Sandstone, with the bridge in Kayenta beds (fig. 5.15). The road beyond is in upper Kayenta and lower Navajo Formations.

Figure 5.15. Eagle Canyon as seen westward from the bridge at Mile 36.6. Wingate Sandstone holds up the cliff in the inner gorge and is capped by flaggy to irregular bedded Kayenta Sandstone in the brushy slope zone. Navajo Sandstone rises on the skyline as the series of white cliffs.

38.6 Beginning of double road cuts through upper, cross-bedded, light yellow gray Navajo Sandstone and overlying well-bedded lower Carmel Formation. Navajo Sandstone is considered to be a broad sheet of eolian sandstone, in part deposited along the southeastern shore of the Carmel Sea. Lower beds of the Carmel Formation are dolomitized but locally contain fair molluscs, particularly near the thin reddish unit in the gray shale, above the dolomite ledges. Upper beds in the road cut, east of the Moore Ranch Road exit, contain thick beds of massive gypsum interbedded with sandy thin-bedded dolomite and siltstone. Carmel rocks record a relatively rapid invasion of the Carmel Sea, and then a gradual oscillating withdrawal.

41.1 Side road east to overlook (fig. 5.16). Turn-around of overlook is on lower Carmel Beds. Thick, cross-bedded, light-colored Navajo Sandstone forms the cliffs in canyons to the east and southeast. Rocks at

Figure 5.16. View southwestward from the rest area at approximately Mile 41.0 Massive exposures in the bottom of the canyon are upper beds of the Navajo Sandstone and are overlain by the well-bedded, fossiliferous lower beds of the Carmel Formation at the canyon rim. Fish Lake Plateau is in the far distance to the southwest.

the road junction are greenish upper Carmel Formation, now dipping moderately steeply off the west flank of the San Rafael Swell. From near the junction a panorama of the brightly colored Jurassic (fig. 5.17) and more somber Cretaceous and Tertiary rocks spreads virtually from horizon to horizon to the west. The road descends down a dip slope held up by the upper Carmel Beds. On the skyline to the west Tertiary rocks cap the Wasatch and Fish Lake Plateaus to the northwest and southwest, respectively. Draped gypsum has flowed into gullies north and south of the road and appears almost as frosting.

44.4 Approximate contact of the gypsiferous red and green Carmel Formation

Figure 5.17 View westward of upper Jurassic and Cretaceous rocks on the west side of the San Rafael Swell and east front of the Wasatch Plateau, taken from approximately Mile 45.0. Light-colored Curtis Sandstone forms the prominent escarpment in the foreground, above slope zones and low ledges on the Entrada Formation. Summerville beds are exposed as the first prominent dark slope above the light- colored Curtis Formation and occur below the banded Morrison Formation. Prominent slopes and the flat country in the middle distance are on Mancos Shale, the lower unit of which is capped by Ferron Sandstone which holds up the cuesta in front of the prominent escarpment of the Wasatch Plateau. Light uppermost Mancos Shale is overlain by the prominent cliff of Star Point Sandstone and younger rocks on the skyline (photograph by W.K. Hamblin).

Figure 5.18 Flat-bottomed sandstone lense in the lower Entrada Formation in road cuts at Mile 47.0. This sandstone is one of several similar accumulations that are thought to be samll bars or splay sandstones in the dominantly tidal flat-marginal marine Entrada Formation.

with the overlying more reddish orange Entrada Formation. The Entrada Formation weathers to "stone babies" and a variety of columns and pillars. Entrada Formation is thought to be a supratidal flat sequence between deposits of the Carmel and Curtis Seas.

47.0 Double road cuts in complexly channeled and lensing Entrada Formation (fig. 5.18). Lacy networks through even the thickest beds are crystalline gypsum.

48.5 Deep double road cut through greenish well-bedded Curtis Sandstone. "Stone baby" beds of upper Entrada weather to form goblins and dragons beneath the Curtis Sandstone on the east face of the escarpment. A side road to a rest area at the west end of the deep cut leads to an overlook area in the upper Curtis Sandstone (fig. 5.19). To the east the valley is lined with columns and pillars in the Entrada Formation. A pavement of a barlike sandstone in the Entrada Formation forms the strongly-jointed shelf below the overlook. West of the overlook "castle walls" of reddish Summerville Formation rise above the Curtis Sandstone cuesta (fig. 5.20).

49.4 Deep road cuts through channelled and rippled reddish Summerville Formation, particularly on the westbound lanes, show excellent tidal flat deposits. Pinkish sandstone shows climbing ripple marks, bimodal current directions, and trace fossils. Soft-sediment deformed structures show in darker reddish brown mudstone beds. Exhumed channel deposits armour ridges to the south of the road and mark locations of paleo-

Figure 5.19. View southward along exposures of the Curtis Sandstone from the rest area at approximately Mile 49.0. The Curtis Sandstone is a sheetlike marine sandstone that marks a minor invasion of the Sundance Sea southward into the Colorado Plateau. The formation occurs as a distinctive greenish marker between the reddish tidal flat deposits of the underlying Entrada Formation and the overlying Summerville Formation.

Figure 5.20 Exposures of upper Summerville Formation, in the foreground, are capped by massive gypsum (G) with Morrison beds (M) along the skyline, as seen westward from approximately Mile 50.0. The gypsum deposits record final drying up of the tidal flat sequence.

channels in the middle part of the formation. Soft sediment has been eroded away from the channels leaving them as linear protective ridge caps.

50.3 Crest of cuesta. Massive light beds to the north in the top of the Summerville Formation are massive gypsum, overlain unconformably by basal sandstone of the Morrison Formation, here somewhat thinner than on the east flank of the dome. Well-bedded maroon and gray shale of the Brushy Basin Member of the formation is exposed above the basal sandy section on the backslope of the cuesta near the road. Several trails lead to uranium prospects within the Morrison Formation northwest of the cuesta crest.

51.6 Coarse conglomerate of the Buckhorn Member of the Cedar Mountain Formation exposed in road cuts. This conglomerate separates the more maroon and varicolored Morrison Formation from the distinctly more gray and ashy-appearing overlying Cedar Mountain Formation. Both these units form peculiar barren badland topography perhaps related to the expansion-contraction weathering of the volcanic clays that the rocks contain. Such weathering makes it almost impossible for most plants to populate the badlands.

Figure 5.21. Eastward to gray Mancos Shale exposures on the south side of the freeway at approximately Mile 54.0. Ferron Sandstone forms the prominent cliff along the skyline toward the right. Debris fans have locally protected the soft underlying Mancos Shale and have helped produce a variety of irregular erosional features.

52.8 Cross Muddy Creek Bridge. Dakota Sandstone is exposed capping the west-dipping cuesta of Cedar Mountain beds immediately east of the bridge. Mancos Shale is exposed west of the bridge and to the south and north along the strike valley. Terraces are cut across the Mancos Shale and are armoured with bouldery debris from the overlying Ferron Sandstone (fig. 5.21). Uneven heaving of the Mancos Shale makes highway construction over the formation difficult, and produces wavy uneven roads like some stretches ahead.

Figure 5.22. Northwestward to lenticular, well- bedded, deltaic and barrier island sandstones of the Ferron Sandstone as seen across Ivie Creek from Mile 55.6. Light-colored sandstones on the skyline are river deposits. Those in the more well-bedded lower part of the cliff are thought to be marginal marine and barrier island associated sandstone accumulations. Coal is common in the upper part of the sandstone unit.

56.0 Road cuts in westbound lane of top of lower Mancos Shale, overlain by coal-bearing barrier-island and deltaic sandstone and shale of the Ferron Sandstone (fig. 5.22). The sandstone tongue here is considerably thicker and coarser grained than equivalent beds on the east side of the San Rafael Swell or where the sandstone is exposed east of Wellington, north of the Swell.

The upward coarsening sequence of marine sparingly fossiliferous siltstone, fine-grained burrowed sandstone and coarse sandstone with channels which are overlain by moderately thick coal beds record a major regression of marine deposits in front of a lobe of a Ferron Sandstone delta. Upper deltaic plain sandstone, lacustrine siltstone, and coal deposits are exposed in road cuts of the upper part of the sandstone, on to the southwest.

56.7 Major channel-fill sandstone near the top of the Ferron Sandstone. Top of the sandstone and base of the fossiliferous middle Mancos Shale are exposed in cuts 0.4 miles to the west. Middle Mancos Shale was deposited over Ferron sediments as the Cretaceous seaway again spread westward over the delta.

58.0 Sevier County-Emery County Line. Road is in Mancos Shale below a pediment surface armoured with basaltic andesite boulders derived from the Fish Lake Plateau, to the southwest. The road rises onto the pediment surface and then descends westward in Mancos Shale.

60.8 Bridge on Price-Emery Interchange at Junction of Utah State Highway 10. Terraces along creeks to the north are cut in Mancos Shale and are andesite-boulder protected. State Highway 10, to the north, follows the "race track" along the easily eroded Mancos Shale around the western margin of the San Rafael Swell and connects to U.S. Highway 50-6 at Price. Interstate Highway 70 continues westward in middle Mancos Shale.

62.8 Fremont Junction. Utah State Highway 72 leads south to Loa and connects with Utah State Highway 24 which crosses through Capitol Reef National Monument.

Emery Sandstone is exposed north of the road, northwest of the junction (figs. 5.23, 5.24). Three or four pulses of barrier-island regressive sandstone show upward coarsen-

EASTERN WASATCH PLATEAU CASTLE VALLEY AREA
FERRON - HUNTINGTON - GORDON CREEK

Period	Formation	Thickness	Remarks
JURASSIC	Morrison Fm	100–400	
	Summerville Fm	260–460	Twist Gulch Mbr. of Arapien Shale
	Curtis Fm	200–300	
	Entrada Ss	850–1100	
	Carmel Fm	1200	= Twelvemile Cyn Member of Arapien Shale
	Navajo Ss	300–500	
TRIASSIC	Kayenta Fm	0-200	
	Wingate Ss	350–440	
	Chinle Fm: Church Rock M	200-300	
	Chinle Fm: Moss Back M	0-90	
	Moenkopi Formation: upper member	740	
	Moenkopi Formation: Sinbad Memb	110	
	Moenkopi Formation: lower member	340	ripple marked ss, siltstone
PERM	Kaibab Ls	140-230	
	Diamond Fork Ss ("Coconino")	430-550	
₽	Hermosa Fm	200	= Oquirrh Fm
	Molas Fm	240	UNCONFORMITY
MISSISSIPPIAN	Deseret - Humbug Fms	300	
	"Redwall" Dolomite	850	
CAMBRIAN	"Lynch" Dolomite	400	
	"Maxfield" Ls	460	
	Ophir Shale	150	
	Tintic Qtzt	180	
P	"Granite and metamorphics"	–	

Period	Formation	Thickness	Remarks
PALEOCENE	Flagstaff Fm	1000	fresh water ls, sh
—?—	North Horn Formation	1600	
CRETACEOUS	Mesaverde Group: Price River Formation	750	
	Mesaverde Group: Castlegate Ss Member	0-200	
	Mesaverde Group: Blackhawk Fm	700–1000	
	Mesaverde Group: Star Point Ss	300	
	Mancos Shale: Masuk Shale Memb	300–1000	
	Mancos Shale: Emery Sandstone Memb	800 1400–	
	Mancos Shale: Blue Gate Shale Memb	1500–2000	blue gray marine sh
	Mancos Shale: Ferron Sandstone Member	300–700	thickens westward
	Mancos Shale: Tununk Sh Member	300–600	
	Dakota Ss	100-175	
	Cedar Mountain Formation	300–700	
	Buckhorn Cg	0-90	

Figure 5.23. Stratigraphic section of rocks exposed along the western side of the San Rafael Swell and in the eastern Wasatch Plateau (from Hintze, 1973).

Figure 5.24. Emery Sandstone, as seen northward from Mile 62.8. The sandstone is in shorefront facies, in front of the barrier island sequence which is developed beneath the Wasatch Plateau to the west. Emery Sandstone is one of the several extensive eastward-spread sheets of sandstone deposited along the western border of the Mancos seaway as a response to rising uplifts in the Sevier orogenic belt to the west.

Figure 5.25. View northward along the eastern escarpment of the Wasatch Plateau. The prominent cliff in the escarpment is the Star Point Sandstone which separates the coal-bearing Blackhawk Formation, above, from the underlying Masuk Shale of the Mancos Shale, below. The Star Point Sandstone continues northward as a prominent cliff along the Wasatch Plateau escarpment as seen in Figure 5.17.

ing as shaly beds are overlain by tan sandstone. Emery Sandstone separates upper and middle Mancos Shale.

63.9 Entering Fish Lake National Forest, at the eastern base of the Wasatch Plateau to the north, and the Fish Lake Plateau to the south. Cretaceous barrier island sandstones form cliffs to the north, with the Star Point Sandstone as the major light cliff (fig. 5.25). Coals above the Star Point Sandstone have burned and baked overlying rocks to produce pinkish stained areas high on the plateau. Star Point Sandstone is exposed as the cliff in the canyon ahead.

67.9 **Summit** (Elevation 7,900 feet). Bleached sandstone to the north is in the Price River Formation, here still dipping westward off the San Rafael Swell dome.

72.5 Excellent view toward the north of nipple-shaped Mt. Musinia, elevation 10,986, and of the fault along the east side of the Musinia graben (fig. 5.26). A graben is a long, relatively narrow downdropped fault

Figure 5.26. View northward along the east boundary fault of the Musinia graben from approximately Mile 72.0. Rounded slopes on the left are in the North Horn Formation, which have been downdropped against Castlegate Sandstone and Price River beds, the ledge -forming units on the right. The highway in the foreground is approximately along the trace of the boundary fault. Mt. Musinia is the snow-covered peak on the skyline toward the left.

block and here the softer pinkish North Horn Beds have been dropped down in the block. More resistant Price River and Castlegate rocks, above coal-bearing Blackhawk Beds (fig. 5.28), are exposed east of the fault. The trace of the fault is visible ahead where ledges of the older Cretaceous rocks terminate westward against softer rounded hills of North Horn Formation. The highway has been constructed here along the trace of the fault.

75.0 Cross Salina Creek in exposures of pinkish and gray North Horn Beds along the western side of the Musinia graben. Cliffs to the west are of Blackhawk Formation west of the graben boundary fault. The graben boundary faults here have displacements of 1,200 to 1,500 feet. Shale and sandstone west of the fault contain a prolific fossil plant assemblage.

79.1 Excellent exposures of coal-bearing Blackhawk Formation are cut along the freeway. Massive Castlegate Sandstone forms cliffs along the canyon wall. Several coal mines were developed in Salina Canyon in the upper part of the Blackhawk sequence. A narrow-gauge railroad was constructed up the canyon to serve the coal mines and part of the right-of-way is still used for a service road off the highway.

80.0 Railroad tunnels on the south of the canyon (fig. 5.27) are in Castlegate Sandstone. Exposures nearby show complex lensing and sedimentary structures characteristic of braided stream deposits.

West of the tunnels the highway crosses several small faults which form horsts and graben and are best shown by relative offset of the massive cliff-forming Castlegate Sandstone. One of these graben is well shown by offset near Water Hollow where Castlegate Sandstone, on the east, is dropped down against Blackhawk beds on the west. Castlegate Sandstone west of the fault is on the canyon rim. It dips gently westward and is at road level approximately 3 miles ahead on the highway.

Figure 5.27. Tunnel on abandoned narrow gauge railroad through Castlegate Sandstone, at approximately Mile 80.0, as seen from the west. Lower beds of the Price River Formation form the pinion, and juniper-covered slopes above Castlegate exposures. This belt of Castlegate Sandstone has been downdropped along the Water Hollow graben in the western part of the Wasatch Plateau.

84.4 Exposures on both the south and north of the road are in flow-folded or slumped shale and thin sandstone in the Price River Formation. Massive Castlegate Sandstone dips below road level one-half mile to the east. Price River deposits are stream-dominated and are usually complexly channeled and sandstone units show much cross-bedding.

85.2 Junction Gooseberry Recreational Area Side Road with Interstate 70. Price River beds dip westward beneath North Horn beds at approximately the junction. Broad open valley here is in part on landslide lobes from the south and north. Green River Formation caps the hills to the southwest and northwest, above reddish Colton, resistant Flagstaff, and resistant North Horn beds.

88.2 Angular unconformity of North Horn Formation on Cretaceous sandstone and shale (figs. 5.28, 5.29) is exposed in deep road cut on the north. Older Cretaceous shale, appearing like Mancos Shale, and rocks as old as Arapien Shale appear be-

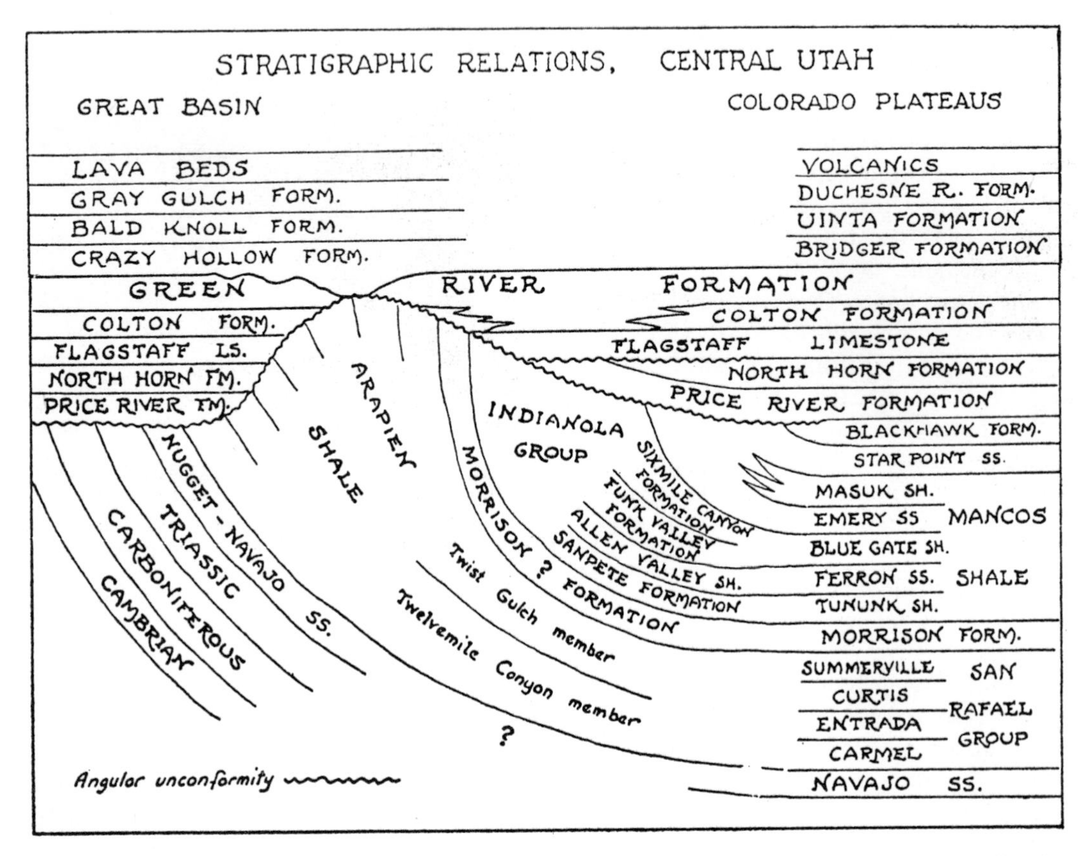

Figure 5.28. Generalized geologic cross section through the western part of the Wasatch Plateau showing the angular unconformities which separate the late Cretaceous and Tertiary beds from the older Cretaceous and Jurassic rocks. The conclusions shown here are those of Spieker who did much of the early geologic mapping and stratigraphic work in the Wasatch Plateau (from Spieker, 1949).

neath the unconformity to the west. Arapien Shale up to Morrison Beds are nearly vertical beneath flat-lying Tertiary beds in the lower walls of the canyon approximately 1 mile to the west.

89.5 Leaving Fish Lake National Forest. Jurassic Arapien Shale (figs. 5.23, 5.30) is exposed both to the north and south near old calcite mill workings.

91.9 Mouth of Salina Canyon. Excellent exposures of lower red Arapien Shale overlain by younger Gray Gulch Formation and volcanic rocks occur in barren badlands on the northern and southern margins of the valley fill at the mouth of the canyon.

93.7 Junction of Utah State Highway 4 Access Road from Interstate 70 with U.S. Highway 89 in downtown Salina. Salina will be bypassed by Interstate 70 when construction is completed. The Interstate Highway skirts around south of town on Arapien Shale and volcanic debris. For a continuation north or south along U.S. Highway 89 see Geologic Guide Segment 2.

Figure 5.29. Angular unconformity exposed in road cuts at Mile 88.2. Older deltaic Flagstaff Limestone above rests on North Horn Formation beds below.

Figure 5.30. Barren exposures of Jurassic Arapien Shale as seen in the mouth of Salina Canyon at approximately Mile 90.5. These rocks are gypsiferous and have a peculiar endemic selenium- and salt-tolerant flora.

Segment 6

0.0 Junction of U.S. Highway 163 with Interstate 70 at Crescent Junction. Turn south onto U.S. Highway 60 toward Moab (fig. 6.1). Low hills to the southwest and southeast are in Mancos Shale, but on the skyline to the south are Jurassic rocks exposed on the upthrown block of the Moab Fault. At ten o'clock folded Jurassic rocks form both flanks of the Salt Valley Anticline (fig. 6.2).

1.1 Approximate axis of Salt Valley Anticline crosses highway. Tertiary intrusions core the La Sal Mountains which rise to approximately 12,000 feet at ten o'clock. Salt Valley Anticline is one of a series of salt intrusion structures developed along the flank of the Paradox Basin. Several of these have graben faults along their crests. Salt Valley Anticline has Cretaceous Mancos Shale (fig. 6.3) dropped down as a "keystone" between flanks of older Jurassic Entrada, Morrison, and lower Cretaceous Burro Canyon Formations. At nine o'clock the Uncompahgre Uplift forms the faint hills on the skyline. The Henry Mountains, at three o'clock, are also cored by Tertiary intrusions and are the locality where Gilbert visualized laccolithic-type intrusions.

4.1 Dakota Sandstone exposed in bluffs east of the road and the spur line of the D&RGW Railroad. The highway is on Mancos Shale above Dakota Sandstone cuestas. Older Burro Canyon beds (fig. 6.3) are visible below the Dakota Sandstone in gullies through the cuesta.

5.4 Cross Rockhouse Creek. Morrison and Burro Canyon beds visible on the flank of Salt Valley Anticline. The road continues in Mancos Shale. A low cuesta to the southwest of the road is held up by silty equivalents to the Ferron Sandstone (fig. 6.4) which is part of a delta complex exposed near the Wasatch Plateau far to the west.

9.5 Side road to microwave relay station to west. At ten o'clock the prominent castellate erosion features in the interior of the Salt Valley Anticline, in Arches National Park, are in the Moab Tongue of the Entrada Sandstone (fig. 6.3). The overlying light-colored Curtis Sandstone and Morrison Formation form the biscuitlike dip slopes.

13.2 Entrance Road to Canyonlands Airport.

13.6 Cross spur line of railroad on overpass. Prominent butte ahead is capped with Dakota Sandstone above slope zone on Morrison and Burro Canyon Formation above Entrada Sandstone cliff.

14.9 Cross down through cuesta cap of Dakota Sandstone onto greenish and varicolored beds of the Burro Canyon Formation.

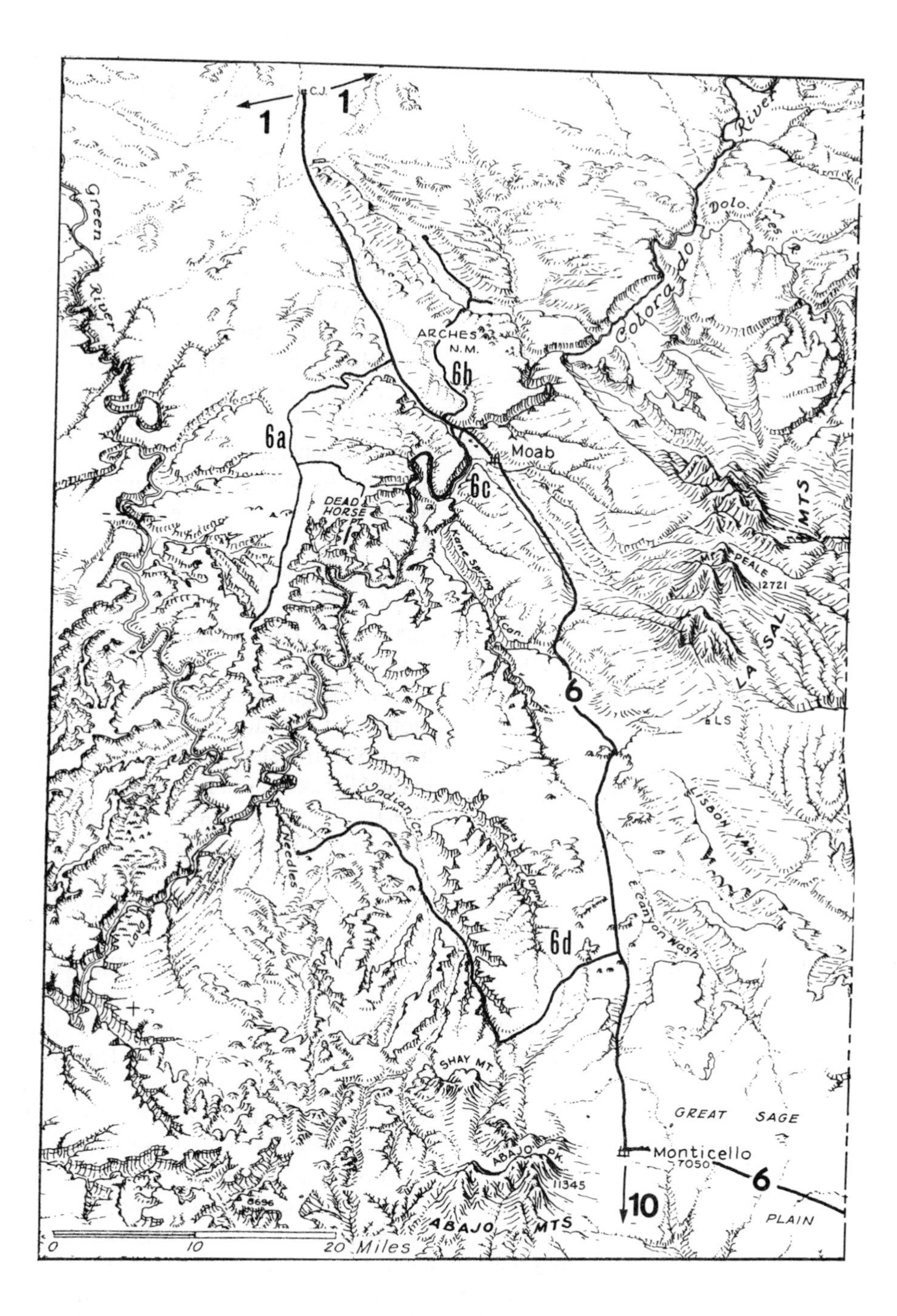

Figure 6.1 Index map of Route 6 from Crescent Junction southeastward to Arches National Park and Canyonlands National Park and Monticello. Route 1 leads westward from Crescent Junction along U.S. Highway 50 and 6 and Route 10 leads southward into the Monument Valley area from Monticello.

Figure 6.2. Southeastward along Salt Valley Anticline to the La Sal Mountains, in the distance, from south of Crescent Junction at approximately Mile 1.1. Rocks from the Jurassic Entrada Formation up through the Morrison Formation are exposed along the flanks of the structure.

15.8 Cross concrete bridge over Courthouse Wash. Greenish beds are lower Cretaceous Burro Canyon Formation. The deep cut to the east is for diversion of the wash which experiences common summer flash floods. Maroon and greenish gray Morrison beds are exposed in badlands along either side of the road beyond the mouth of the gully.

17.9 Bright green Burro Canyon beds are faulted against cliff-forming Wingate and underlying Chinle and Moenkopi beds at three o'clock to the southwest (fig. 6.5). To the north Morrison beds are exposed on the stripped surface on the south side of Salt Valley Anticline and north flank of Courthouse Wash Syncline.

18.8 Road is in massive sandstone of Morrison Formation. To the southwest bright green beds of the Chinle Formation are exposed to uranium mine access road at about the base of the formation and the top of the Moenkopi beds. Top of the cliff above is in Kayenta beds above the prominent cliff of Wingate Sandstone. The trace of the Moab Valley Fault is close against the base of the escarpment where reddish Cutler and Moenkopi beds form irregular erosional features (fig. 6.6).

20.3 Side Road to Canyonlands National Park and Dead Horse Point State Park to the south. See Guide Segment 6A for log of this road. Morrison beds on both sides of the road are downfaulted against Permian Cutler redbeds to the south along the Moab Fault. U.S. Highway 163 crosses the trace of the Moab Fault approximately 0.7 mile ahead to the southeast.

22.4 Overlook into deep railroad cut in Cutler redbeds. The railroad was constructed to service potash mines developed along the Colorado River downstream from Moab.

24.6 Shallow bend in the highway. Morrison beds on the north are faulted down against the pink to maroon arkose of the Cutler sequence.

25.6 Begin bend on highway near the mouth of Moab Canyon. Light-colored limestone and shale in cuts to the south are in the Honaker Trail Formation, as are those directly ahead. Marine fossils occur in the shale and limestone north and east of the railroad fill by the deep double railroad cut.

At the north end of the cliff on the bend, the highway crosses the Moab Fault (fig. 6.7). Here massive Jurassic Entrada Sandstone is faulted against Pennsylvanian-Permian Honaker Trail and Cutler (fig. 6.8) rocks. The crinkled maroon Dewey Bridge Member of Entrada forms the erosional break between the massive Entrada Sandstone above and the fractured lighter colored Navajo Sandstone below.

26.2 Junction of Side Road into Arches National Park with U.S. Highway 163. Park headquarters and Visitors Center are visible a short distance down the side road to the northwest, at the base of the cliffs of En-

GREEN RIVER AREA, UTAH

Subsurface thicknesses from well data.

System	Series	Group	Formation / Member	Thickness (ft)	Notes
JURASSIC		San Rafael Group	Entrada Ss	520	Permo-Jurassic wind directions- Poole (1962)
			Carmel Fm	120	
		Glen Canyon Group	Navajo Ss	450	
			Kayenta Fm	170-270	
			Wingate Ss	300—400	
TRIASSIC			Chinle Fm	240-300	Chinle regional: Stewart (1969)
			Moss Back Cg M	40-140	
		Moenkopi Fm	upper memb	500	
			Sinbad Ls M	50	
			lower memb	150	
PERMIAN	Leon		Kaibab Limestone	60-160	
		Cutler Group	White Rim Ss	300	Coconino of some well logs
	Wolfcampian		Organ Rk Sh	0-200	
			Cedar Mesa Ss	200-400	
			Elephant Canyon Formation	600	*Pseudoschwagerina Dunbarinella Triticites Schwagerina* "Rico Fm" of older reports
PENNSYLVANIAN	Missourian	Hermosa Group	Honaker Trail Fm	800—1200	
	Desmoinsian		Paradox FM	0—3000	salt
	M-A		Molas Fm	50	
MISS			Redwall Ls (Leadville of Colo.)	500—700	*endothyrid foraminifera, corals, crinoid columnals*
DEV			Ouray Ls	100	
			Elbert Fm	230	
			McCracken Ss M	0-40	UNCONFORMITY
CAMBRIAN			Cambrian limestones undifferentiated	800—1000	"Lynch" Ls; "Maxfield" Ls
			Ophir Fm	200	
			Tintic Qtzt	160-200	
P€			Precambrian granite	—	1.8 B.Y. Rb-Sr

System	Group	Formation	Member / Facies	Thickness (ft)	Notes
EOCENE		Green River Fm		0—7000	oil shale fresh water ls and shale
PALEOCENE	"WASATCH GROUP"	Colton Fm ("Wasatch" of some authors)		1500	reddish continental mudstone, siltstone, and sandstone
		Flagstaff Ls		0-100	fresh water ls
		North Horn Fm (Tuscher Fm)		400-500	UNCONFORMITY
CRETACEOUS	"MESAVERDE GROUP"	Price River Formation	Farrer facies (generally noncoal-bearing)	725	
			Neslen facies (coal bearing)	200-250	1' coal
			Sego Ss M	50-150	
			Buck Sh Tong	0-50	Mancos tongue
			Castlegate Ss M	85	two 3' coals
		Black Hawk	Deseret Memb	140	6' coal
			Grassy Memb	200-250	pinches out to east
			Mancos Sh Ton	100-150	
			Sunnyside M	200-0	*Nucula*
		Mancos Shale	Pierre fauna; Telegraph Creek fauna; Niobrara fauna	3000	*Scaphites hippo*; *Inoceramus*; *Inoceramus lobatus*; *Baculites aquilaensis*; *Desmoscaphites bass*
			Ferron Ss M	10	*Inoceramus Ostrea Baculites Collignoniceras*
			Tununk Shale Member	800	*Ostrea (Carlile age)*
		Dakota Ss		0-40	*Gryphaea newberryi*
		Cedar Mtn Fm		250	Buckhorn Cg Mbr
JURASSIC		Morrison Fm	Brushy Basin	300	green and purple dinosaur beds
			Salt Wash M	180	
	San Rafael Group cont'd		Summerville	20-130	
			Curtis Fm	0-75	chocolate beds
			Entrada Ss	(520)	Arches Nat'l Mon

Figure 6.3. Stratigraphic section of formations in the Green River Area (from Hintze, 1973).

Figure 6.4. Northwestward along the gentle cuesta capped by silty beds of the Ferron Sandstone at their easternmost extent in the Mancos Shale, from approximately Mile 9.5.

Figure 6.6. Prominent cliffs of Wingate Sandstone are capped by moderately well-bedded Kayenta Sandstone and occur above slopes of Chinle Formation. These exposures are on the east side of the side road which leads south into Canyonlands National Park, at approximately Mile 20.3. Moenkopi beds form the lowest part of the slope above the Cutler cliffs, which are exposed in the foreground near the bottom of the canyon wall.

Figure 6.5. Cliff-forming Wingate Sandstone, capped by moderately well-bedded Kayenta Sandstone, on the left, is offset along the Moab Valley Fault against slope-forming Morrison and Cedar Mountain Formations on the right as seen from approximately Mile 18.0.

Figure 6.7. View southwestward over the Visitors Center to the narrows at the lower end of Moab Canyon, from the park above the headquarters of Arches National Park. Cliffs on the skyline are in Wingate Sandstone, above slopes of Moenkopi and Chinle beds. Pennsylvanian and Permian rocks are exposed in narrow road cuts in the mouth of the canyon and in the deep railroad cuts to the left. The Moab Fault crosses the highway at the mouth of the canyon and has dropped Navajo Sandstone, Carmel Formation, and massive Entrada Sandstone down against the upper part of the Paleozoic sequence near the prominent bend on the park road, to the right of the canyon mouth.

MOAB - LA SAL AREA

System	Group	Formation	Thickness	Remarks
PENNSYLVANIAN	HERMOSA GROUP	Honaker Trail Fm	1000–2500	"Rico" facies 0-500' interbedded ls, sh, arkose; *many fossils Composita Productus Triticites Chonetes Polypora*
		Paradox Fm (cyclic sequence of evaporites, black shales, sandstone, limestone, and locally arkose.)	2000–7000	Paradox and earlier strata are known here chiefly from geophysical and well data; salt, bedded and intrusive masses; potash
		Pinkerton Trail Fm	0–200	*Anthracospirifer*
		Molas Fm	0-50	
MISS		Leadville Ls (Redwall) upper	150–400	oil producing zones- Big Flat and Lisbon oil fields
		Leadville Ls (Redwall) lower	100–300	
DEV		Ouray Ls	0-175	
		Elbert Fm	100-200	
		McCracken Ss M	25-100	
CAMBRIAN		"Lynch" Dolomite	400	
		Ignacio Fm	500–900	*Obulus*
P-Є		Precambrian schist, gneiss, quartz monzonite	—	quartz monzonite 1480 M.Y. gneiss- 1700 M.Y.

System	Formation	Member	Thickness	Remarks
QUAT	Gold Basin Fm, Beaver Basin Fm, Placer Creek Fm, Harpole Mesa Fm		0–few hundred	till, alluvial sand and gravel, eolian sand
PLIO	Geyser Creek Fanglomerate		0-600	
OLIGOCENE	syenite porphyry dike swarm			
	monzonite porphyry dikes and sills			
	diorite porphyry laccoliths			24 M.Y. U-Pb (Armstrong, 1969); intrusive contact
CRET	Mancos Shale	Blue Gate Sh	0-400+	
		Ferron Ss M	0-60	
		Tununk Shale M	800	*Gryphaea*
	Dakota Ss		100-200	ss, coal, shale
	Burro Canyon Fm		80-250	Cedar Mtn Fm
JURASSIC	Morrison Fm	Brushy Basin Sh Member	250–400	*dinosaur bones*
		Salt Wash Ss Memb	200–350	
	Summerville Fm		20-80	
	Entrada Ss	Moab Member	50-180	white sandstone
		Slick Rock Member	240–360	Arches Nat'l Mon
		Dewey Bridge M	60-180	
	Navajo Sandstone		0–400	thins eastward; Permo-Jurassic wind directions: Poole (1962)
TRIASSIC	Kayenta Fm		140–260	
	Wingate Ss		150–300	vertical cliff
	Chinle Fm	Church Rock Member	260-350	red sandy facies see Stewart (1969)
		Owl Rock M	0-190	
		Moss Back M	0-90	greenish silt here, few pebbles
	Moenkopi Fm		0-400	reddish brown
PERMIAN	Cutler Fm	White Rim Ss M	0-250	not present east of Dead Horse Pt, see Baars and Seager (1970)
	Cutler Fm (arkosic facies)		400-3000	maroon and purple arkose, cg, and reddish brown sandy mudstone

Figure 6.8. Stratigraphic section of rocks of the Moab-La Sal Area (from Hintze, 1973). Lower Paleozoic rocks are known only from subsurface information and are not exposed in the canyonlands area.

trada and Navajo Sandstone. See Guide Segment 6B for a log of the road into various parts of Arches National Monument.

27.2 Junction of Utah State Highway 279 with U.S. Highway 163. Utah State Highway 279 leads south along the western side of a uranium mill tailings pond and then south into the gorge of the Colorado River to a potash plant development. See Guide Segment 6C for a log of this route. Moab Valley, into which we've now entered, is a salt anticline that is en echelon with Salt Valley. Moab Valley also has a collapsed "keystone" of Mancos Shale and older rocks apparently produced by collapse following solution of salt at depth. Faulted Wingate and Chinle rocks are exposed north of Atlas Minerals uranium concentration plant, and on the southwest margin of Moab Valley.

28.2 Cross Courthouse Wash. Uppermost Chinle beds exposed in road cuts both east and west of the bridge show northward dip off the Moab Valley salt anticline into Courthouse Wash Syncline.

28.8 South Bridge Abutment Over Colorado River and Junction of Utah State Highway 128 with U.S. Highway 163, the state highway which leads northward upstream along the river to Cisco and Castle Valley. Upper beds of Chinle Formation and massive cliffs of Wingate Sandstone and beehive-weathering Navajo Sandstone are visible upstream and downstream to the south near where the Colorado River exits from the valley. Paradox Formation piercement features are present east of the river gorge on both the northeast and southwest margins of the valley.

30.0 Enter Moab. Chaotic Paradox gypsiferous units are exposed along the border fault zone of the valley to the north.

31.4 Bridge over Mill Creek at the south end of Moab business district. Tan and gray Paradox piercement bodies are visible through the trees to the southwest along the southern border fault of the collapsed "keystone." Jointed Wingate and Navajo Sandstones cap the faultline escarpment. Peaks of the La Sal Mountains rise above the Jurassic and Triassic rocks to the east.

38.0 Grand County-San Juan County Line. Rolling country nearby is in Morrison Formation dipping beneath gray Mancos Shale of the collapsed "keystone" to the east. Older Jurassic rocks form the valley walls beyond the border faults.

43.2 Begin climb out of Moab Valley across badly fractured Morrison beds which are here fault juxtaposed against Chinle and Wingate Formations. The road continues over the summit in Morrison beds.

46.4 Cane Springs State Park. The springs issue from near the base of the massive Moab Tongue of the Entrada Sandstone near a fault which has dropped Morrison and older beds on the west down against the Entrada Sandstone on the east. The fault is particularly dramatic approximately one-half mile north of the park area where slope-forming Morrison abuts massive cliff-forming Entrada.

South of the park the highway continues in crinkly-bedded Dewey Bridge Member of the Entrada Formation on the tableland at the top of the Navajo Sandstone and below the cliffs of Entrada Sandstone.

47.6 Cross drainage. Road is in Dewey Bridge beds to the south and north, but drops onto upper Navajo near the drainage. Ledge- and slope-weathering Morrison Formation caps the Entrada cliff on the east. The road continues at about this same stratigraphic level for several miles.

51.6 Natural Gas Compressor Station on the west. Crinkly maroon Dewey Bridge

Member is exposed at the base of the cliff of beehivelike outcrops of massive Entrada Sandstone. Navajo Sandstone is exposed in gullies to the west.

53.3 Junction of Utah State Highway 46 with U.S. Highway 163. The state highway leads east to La Sal and the Big Indian uranium district. Massive sandstone east of the junction is Entrada Sandstone beneath intensively prospected uranium-bearing Morrison Formation. Mt. Peale, the high southern peak in the La Sal Mountains to the northeast, is cored by a Tertiary intrusion and has an elevation of 12,721 feet.

56.6 Window Arch Turnout (fig. 6.9). The road is in Dewey Bridge beds and the arch is in Entrada Sandstone. Navajo Sandstone is exposed in the gully to the west. Abajo Mountains, held up by Tertiary igneous intrusons, are visible on the skyline to the south from near here and are part of the distinctive igneous province of this part of the Colorado Plateau.

Figure 6.9. Window Arch as seen from the north at approximately Mile 56.5. The arch is in the Moab Member and the road is on the Dewey Bridge Member of the Entrada Sandstone.

For the next seven miles the highway is constructed at approximately the level of the thin red crinkled Dewey Bridge Member, with some road cuts exposing upper Navajo Sandstone and some others the lower part of the massive Entrada Sandstone which generally forms cliffs east of the road (fig. 6.10).

63.9 Side Road to Canyon Rim Recreation Area-Needles Overlook. The road to the west leads to scenic overlooks on the rim of the Colorado River gorge, across from Dead Horse Point State Park. To the east massive Jurassic rocks are dipping westward off the Lisbon Valley Anticline, another salt-related fold. The Lisbon Valley oil field is developed along the cuesta to the southeast, on the west flank of the surface anticline, but in a subsurface anticline in Mississippian Redwall Limestone.

Figure 6.10. View northward of rounded, massive, sandstone cliffs of the Entrada Formation overlain by slope zones on the Summerville and Morrison Formation, as seen northward from approximately Mile 60.0. The road and the slope zone in the foreground is on the crinkly Dewey Bridge Member of the Entrada Formation, above massive white Navajo Sandstone which is exposed in the gullies on either side of the road.

70.7 Junction of Utah State Highway 211 with U.S. Highway 163. The state highway leads west to Indian Creek State Park and the eastern part of **Canyonlands Na-**

tional Park. For a description of the state highway route and access roads in the park see Geologic Guide Segment 6D.

Church Rock (fig. 6.11), a small Entrada outlier to the east of the junction, gives the valley its name. Low bluffs to the east are also in Entrada Sandstone, overlain by slope-forming Summerville and Morrison Formations.

Figure 6.11. Church Rock, an outlier of the Entrada Formation in the center of Church Rock Valley, as seen eastward from approximately Mile 71.0 near the junction of the highway that leads westward into Canyonlands National Park.

74.2 Top of the Entrada Sandstone is at valley level. Summerville Formation is exposed in poor outcrops along the highway and in vertical castellate outcrops along the western wall of the valley. The formation is only 25 to 30 feet thick here and is overlain by the more sandy, stream-deposited Morrison Formation on both sides of the valley. Small faults locally offset the distinctive formation bands (fig. 6.12).

75.3 Begin dugway through relatively poor exposures of Morrison Formation. Dakota Sandstone forms the rim at the crest of the canyon walls.

77.4 Cross small drainage. Massive

Figure 6.12. Westward from the highway at approximately Mile 74.2 to a small fault which has offset the light-colored Entrada Sandstone. A thin deep reddish Summerville Formation overlies the massive Entrada beds and occurs below the prominently bedded Morrison Formation.

sandstone south of the bridge is Dakota Sandstone whose basal contact is approximately at the bridge level.

The road climbs up through the Dakota Sandstone for the next mile and rises out onto a rolling stripped surface on top of the formation at the head of the small canyon. The intrusion-cored Abajo Mountains rise above the stripped plain and are flanked by their own debris.

80.6 Entrance to Monticello airport on the west. Mancos Shale is exposed in road cuts from near here into Monticello. Some of the shale exposures are armoured with gravel swept down from the Abajo Mountains to the southwest.

84.9 Junction of U.S. Highway 666 with U.S. Highway 163 in Monticello. Follow Guide Segment 6 for a continuation of the guide along U.S. Highway 666 toward Cortez and Durango and follow Guide Segment 7 for a continuation of the route log toward Blanding, Bluff, and Monument Valley and across the Navajo Indian Reservation toward Grand Canyon National Park.

85.3 Leaving Monticello east on U.S. Highway 666. The road for several miles is constructed in the lowermost Mancos Shale across the flat uplands and in uppermost Dakota Sandstone across inscribed drainages. Much of the Great Sage Plain, through which we are now driving, is veneered with reddish brown loess, windblown dust and silt, which has helped make the area a moderately fertile farming region.

88.3 Cross Vega Wash. Excellent exposures of Dakota Sandstone can be seen both to the south and north. Lower Mancos Shale is exposed in road cuts both east and west of the wash.

92.1 Double road cuts in Mancos Shale, with weathered reddish loess blanket on top, are typical of the next 5 miles. Juniper woods are left on some of the less favorable areas.

96.5 Junction of side road south to Eastland, U.S. Highway 666 continues in lower Mancos Shale.

101.2 Colorado-Utah State Line. To the east the highway crosses through poor exposures of Dakota Sandstone in the divide areas and good exposures in the canyons and gullies for the next several miles.

106.7 Junction Colorado State Highway 141 North to Egnar. Dakota Sandstone is exposed in gully bottoms wherever topography dips below the general base of the Mancos Shale and where red loess of Dove Creek soil has been removed by erosion.

107.5 Pass three tall concrete elevators west of Dove Creek.

107.9 Entering Dove Creek, the pinto bean capital of the world according to local billing.

108.8 Leaving Dove Creek heading southeast on U.S. Highway 666. Dakota Sandstone is exposed in gully banks and bottoms beneath the reddish loess soil.

112.2 Small strip mines in Dakota Sandstone on the north. Coal in the formation appears to be as much as 3 feet thick here and to occur in a deltaic or low barrier-island complex.

114.5 Cross Sharp Creek. Extensive exposures of Dakota Sandstone are visible along the creek and are capped, on the west side by Mancos Shale above upper Dakota and Carbonaceous shale. Similar exposures can be studied in Alkali Creek, approximately 1 mile ahead to the southeast.

116.7 Large elevators in **Cahone.** Dakota Sandstone is exposed all through town.

118.4 Dolores County—Montezuma County Line. U.S. Highway 666 continues through poor exposures of Dakota Sandstone. The Abajo Mountains to the west, La Sal Mountains to the north, and Ute Mountain to the south are all cored with late Tertiary igneous intrusions that are relatively fine grained. Apparently the intrusions were relatively cold when they were implaced.

121.5 Pleasantview community.

125.5 Yellowjacket, a single-store farming community. Deep red Dove Creek soil blankets Dakota Sandstone. The loess soil is up to 8 or 9 feet deep in some areas, even on crests of some of the low hills. Cross-bedded white sandstone of the Dakota Formation is visible in gullies to the west.

129.7 Junction of Colorado State Highway 147 with U.S. Highway 666. State Highway 147 leads east to Dolores and the Dolores River Valley. Mancos Shale is exposed here in broad downward-bowed structure below surrounding Dakota Sandstone exposures along the highway to the southeast and northwest. Dakota Sandstone is exposed in small canyons and gullies and locally has been prospected for coal.

130.1 Double road cuts in Dakota Sandstone and Mancos Shale where highway passes beneath large irrigation flume. Mancos Shale exposed on the divides and Dakota Sandstone is exposed in the gullies from here to Cortez. Morrison Formation is exposed in some of the deep canyons some distance west of the highway. Dove Creek loess soil covers much of the uplands.

135.3 Well-exposed coally Dakota Sandstone in road cuts and gully exposures.

137.2 Cross-bedded Dakota Sandstone shows well on the north side of the road. The softer coally parts of the formation are less well exposed but are evident in areas where the formation has been prospected or mined on a local basis, particularly in the northwestern outskirts of Cortez.

138.4 Triangle Separation of U.S. Highways 160 and 666 in Western Cortez. U.S. Highway 666 continues south toward Shiprock, New Mexico and U.S. Highway 160 continues east to the entrance to Mesa Verde National Park, Durango, and Pagosa Springs. Cortez is built on Dakota Sandstone and Mancos Shale. U.S. Highway 160 east of town is, in large part, on Mancos Shale for some distance at the foot of Mesa Verde (fig. 6.13).

Figure 6.13. View eastward along the northern escarpment of Mesa Verde from near Cortez, Colorado. Mancos Shale forms the flat land and lower gray slopes at the bast of the escarpment. Sandstone beds of the Mesa Verde Group cap the ridge and the plateau surface.

Segment 6A

Dead Horse Point

0.0 Junction of the Dead Horse Point State Park Road with U.S. Highway 160 heading south towards Canyonlands. The road at the junction is in the Morrison Formation, north of the Moab or West Spanish Valley Fault.

0.2 Cross the D&RGW Railroad Tracks. A short distance south of the crossing the road crosses the trace of the Moab Valley Fault. Cliffs on either side of Seven Mile Canyon ahead are in the Wingate Sandstone with the grayish slope zone below in the Chinle Formation. The Moenkopi Formation and brick red Cutler beds are exposed at the mouth of the canyon (fig. 6.14).

1.2 Moenkopi beds exposed to the west (right) of the road.

1.5 Basal beds of the Chinle Formation rest on the Moenkopi Formation in exposures to the west (right).

1.8 The Moss Back Member of the Chinle Formation is exposed at road level on the west side of the canyon but some distance above the valley on the east side.

2.3 Reentrant in the Wingate cliff where the base of the Wingate Sandstone is at road level on top of reddish flaggy sandstones of the Chinle Formation.

2.7 Canyon junction, the road bends up tributary to the right. Massive ledges of Wingate Sandstone at river level on both sides of the canyon. Slabby Kayenta Sandstone is exposed above the Wingate beds on both sides.

Figure 6.14. View eastward to the cliffs and slopes of Permian and Triassic rocks at the mouth of the canyon at Mile 1.0. Strongly-jointed cliffs that cap the slope zone are in Wingate Formation. The slightly less well-developed cliff below the Chinle slope is the Moss Back Member of the Chinle Formation. Slope zones on well-bedded rocks below are, in large part, in the Moenkopi Formation above Cutler beds.

3.3 Cattle Guard and Bridge Over Canyon. Beginning of climb up the south wall of the canyon through the Kayenta Sandstone. The top of massive ledges of the Wingate Sandstone give way to flaggy beds of the Kayenta Sandstone.

Figure 6.15. View northward to the Monitor and Merimac Buttes fron Mile 3.8. Cliffs of the two buttes are Entrada Sandstone and are capped with thin Summerville and Morrison Formation. The upper part of the bedded sandstone sequence on the spur below is in the resistant Navajo Sandstone. Lower juniper-covered beds are in the Kayenta Formation.

3.8 Monitor and Merrimac Buttes are the large square-topped buttes behind to the northwest (fig. 6.15). Submarine Point is in the Navajo Sandstone below and to the southwest of Monitor Butte. Both buttes expose sections of Dewey Bridge and Moab Members of the Entrada Formation and Curtis, Summerville, and Morrison Formations (fig. 6.16).

4.0 Switchback in the upper part of the Kayenta Sandstone.

4.3 Massive cross-bedded Navajo Sandstone forms rounded beehive-shaped hills on both sides of the road.

5.4 Cross-bedded Navajo Sandstone beehives on both sides of the road.

6.1 Prominent bluffs to the north are in the massive Moab Member of the Entrada Sandstone and are separated from the underlying cross-bedded windblown Navajo Sandstone by the more easily eroded crinkly Dewey Bridge Member of the Entrada. The white massive sandstone on top of the Entrada Cliff is the Curtis Sandstone. Gentle slopes are carved on the thin brick red Summerville Formation, with massive sandstone and characteristic Morrison Formation capping the butte. The road continues through rolling topography cut in the Navajo Sandstone.

8.5 Junction of primitive road to Spring Canyon to the west, leading into the Bowknot Bend area and to the old primitive access road to Dead Horse Point. Continue south on the Utah State Highway toward Dead Horse Point State Park.

9.5 The road drops down through the Navajo Sandstone onto the stripped surface in the Kayenta. The road for some miles ahead is on this surface.

10.3 Excellent exposures of the upper part of the Kayenta Formation along both sides of the road. Small bluffs in the basal cross-bedded part of the overlying Navajo Sandstone caps hills to the east (left). Occasional glimpses of cross-bedded thin sandstone units in the Kayenta Formation. Most of the upper part of the flat here, and Big Flat to the south, is floored by Kayenta rocks for Navajo has been stripped away.

12.4 Junction of Road to the Green River Down Mineral Canyon. This is the main access road to the western part of the lower White Rim, on an exciting switchback road down the Wingate Cliff.

14.2 Excellent view to the north into the Courthouse Wash Syncline and Salt Valley Anticline area across the King's Bottom Syncline. To the east are the La Sal Mountains and to the southwest are the Henry Mountains on the skyline.

14.7 Junction of the Canyonlands National Park Road with the Dead Horse Point State Road, Turn Left to Dead Horse Point State Park. This is at the entrance to Big Flat, one of the high grassy areas which is grazed when water is available. For a continuation toward Canyonlands National Park see Part 2.

16.5 Road Junction. The primitive road

MOAB - LA SAL AREA

System	Group	Formation	Thickness	Remarks
PENNSYLVANIAN	HERMOSA GROUP	Honaker Trail Fm	1000–2500	"Rico" facies O-500' interbedded ls, sh, arkose; *many fossils Composita Productus Triticites Chonetes Polypora*
		Paradox Fm (cyclic sequence of evaporites, black shales, sandstone, limestone, and locally arkose.)	2000–7000	Paradox and earlier strata are known here chiefly from geophysical and well data; salt, bedded and intrusive masses; potash
		Pinkerton Trail Fm	0–200	*Anthracospirifer*
		Molas Fm	0-50	
MISS		Leadville Ls (Redwall) upper	150–400	
		Leadville Ls (Redwall) lower	100–300	oil producing zones- Big Flat and Lisbon oil fields
DEV		Ouray Ls	0-175	
		Elbert Fm	100-200	
		McCracken Ss M	25-100	
CAMBRIAN		"Lynch" Dolomite	400	
		Ignacio Fm	500–900	*Obulus*
P-Є		Precambrian schist, gneiss, quartz monzonite	–	quartz monzonite 1480 M.Y. gneiss- 1700 M.Y.

System	Formation		Thickness	Remarks
QUAT	Gold Basin Fm, Beaver Basin Fm, Placer Creek Fm, Harpole Mesa Fm		0–few hundred	till, alluvial sand and gravel, eolian sand
PLIO	Geyser Creek Fanglomerate		0-600	
OLIGOCENE	syenite porphyry dike swarm			
	monzonite porphyry dikes and sills			
	diorite porphyry laccoliths			24 M.Y. U-Pb (Armstrong, 1969)
CRET	Mancos Shale	Blue Gate Sh	0-400+	intrusive contact
		Ferron Ss M	0-60	
		Tununk Shale M	800	*Gryphaea*
	Dakota Ss		100-200	ss, coal, shale
JURASSIC	Burro Canyon Fm		80-250	Cedar Mtn Fm
	Morrison Fm	Brushy Basin Sh Member	250–400	*dinosaur bones*
		Salt Wash Ss Memb	200–350	
	Summerville Fm		20-80	
	Entrada Ss	Moab Member	50-180	white sandstone
		Slick Rock Member	240–360	Arches Nat'l Mon
		Dewey Bridge M	60-180	
	Navajo Sandstone		0–400	thins eastward Permo-Jurassic wind directions: Poole (1962)
TRIASSIC	Kayenta Fm		140–260	
	Wingate Ss		150–300	vertical cliff
	Chinle Fm	Church Rock Member	260-350	red sandy facies see Stewart (1969)
		Owl Rock M	0-190	
		Moss Back M	0-90	greenish silt here, few pebbles
	Moenkopi Fm		0-400	reddish brown
PERMIAN	White Rim Ss M		0-250	not present east of Dead Horse Pt. see Baars and Seager (1970)
	Cutler Fm (arkosic facies)		400-3000	maroon and purple arkose, cg, and reddish brown sandy mudstone

Figure 6.16. Stratigraphic section of the rocks exposed in Canyonlands National Park area in the Moab-La Sal area (from Hintze, 1973).

to the east leads over the Potash escarpment. **Continue on the Paved Road to the Southeast to Dead Horse Point State Park.** Abajo Mountains are visible ahead to the southeast. These mountains are cored by Tertiary intrusions somewhat similar to the La Sal Mountains which form the high peaks behind Moab to the east. The high country south from Big Flat is the Island in the Sky section of Canyonlands National Park.

20.2 Park Boundary of Dead Horse Point State Park.

21.3 First view of the Canyonlands to the east.

21.5 Entrance Road to the Visitors Center. Stop and pay fees. There is a one dollar entry fee to the park. Exhibits, some soft drinks, candy, and information on the park are available at the Visitors Center. The Visitors Center is built of ripple-marked sandstone. Many of the slabs also show excellent fucoids, and tracks and burrows. Many of the structures associated with the sandstone are typical of shallow water or tidal flat deposits.

22.2 High point and rise. To the south can be seen a limited view of Canyonlands. The cliffs directly ahead form the eastern promontory of the Island in the Sky section of Canyonlands National Park. Dead Horse Point is the small promontory on which we are now riding. Outcrops on both sides of the road are still in the Kayenta Formation. The massive cliffs immediately below are in the Wingate Sandstone.

22.7 Crossing the Narrow Neck into Dead Horse Point. Outcrops in the immediate vicinity are in the lower part of the Kayenta Formation above the cliff of Wingate Sandstone.

23.0 End of the Road at the Dead Horse Point Campground. Walk to the South to the Observation Point over cross-bedded, windblown, and stream-deposited Kayenta Sandstone. Southwest from the observation point at Dead Horse Point the prominent bend in the river exposes Honaker Trail beds at the base of the first cliff of the prominent ledge-forming zone. This is a resistant abundantly fossiliferous limestone unit. The Elephant Canyon-Cutler sequence rises in the center of the meander above this gray table to form the next prominent red vertical wall. Above this particularly to the southwest, the prominent White Rim Sandstone can be seen forming the shelf below the walls that rise to the Island in the Sky section. Above the White Rim deep redbeds of the Cutler are exposed below the brown slopes of the Moenkopi sequence and the gray green shales of the slope-forming Chinle Shale.

Vertical walls to the west rim the Island in the Sky section of Canyonlands National Park and are in the Wingate Formation, the same unit that forms the vertical walls beneath us here at Dead Horse Point. Top of the vertical wall is capped by flaggy-bedded Kayenta Sandstone, and in the Island in the Sky section white beds of Navajo Sandstone form the rounded cliffs beneath the upland surface. The upland is carved on a stripped surface of the Dewey Bridge Member of the Entrada Formation.

Due south into Canyonlands we can see the prominent development of the White Rim Sandstone and Cedar Mesa Sandstone making up the Maze section and the Needle section of Canyonlands National Park. These units are older Paleozoic rocks rising up on the north flank of the Monument Valley Upwarp.

To the east and southeast, across the canyon, Hatch Point exposes essentially the same beds as in the immediate vicinity. To the east, high prominent gray peaks carved on intrusive masses of the La Sal Mountains show well along the skyline. The related igneous-cored Abajo Mountains can be seen directly south over Hatch Point, east of the Needles part of Canyonlands National Park.

Segment 6A

Part 2

0.0 Junction of Canyonlands National Park Access Roadwith Dead Horse State Park Road (Utah State Highway 313). The road into Canyonlands is a graded unpaved road and leads south from the junction across Big Flat which is on the Kayenta Formation. The La Sal Mountians are visible on the skyline to the east and the Abajo Mountains on the skyline to the south. Both ranges are cored by Tertiary igneous intrusions.

2.4 Outliers of Navajo Sandstone are rounded hills to the west. The gorge of the Colorado River is to the east and is rimmed by vertical cliffs of Wingate Sandstone below the Kayenta Formation which is near road level.

3.3 Side road to the west leads past small rounded outliers of Navajo Sandstone.

3.8 Navajo Sandstone exposed to the west. The view on farther to the west is across the gorge of the Green River and toward the Henry Mountains on the skyline. In the far distance to the northwest the white Reef of the San Rafael Swell and the higher Wasatch Plateau are visible on clear days.

4.7 Entering Canyonlands National Park, Island in The Sky Section. The park road is built across lower Navajo and upper Kayenta Formations.

6.2 Junction of side road to the east at the head of **Shafer Trail.** The upper 2.5 miles of the trail skirts the rim of vertical Wingate Sandstone on a semislope shelf of Kayenta Formation below the cliffs of Navajo Sandstone (fig. 6.17). Beyond that point the road descends over the cliff in a series of steep, sharp switchbacks to connect with a lower road built around the canyon margin on the lower White Rim. The White Rim road con-

Figure 6.17. View northward along the upper part of the Shafer Trail. Shafer Trail leads upward from the state highway at Potash, here climbing over a Chinle slope zone and in switchbacks up across the massive Wingate Sandstone cliffs. The upper shelf along which the road passes is in the Kayenta Formation and uppermost low cliffs at the left are in Navajo Sandstone.

nects to Utah State Highway 279 at Potash but the lower Shafer Trail is a rough road and should not be attempted during wet weather or when storms are threatening. One can turn around at various places in the first 2.8 miles and the exciting rim road offers spectacular views of the country below.

7.3 Island in the Sky Ranger Station. Maps and information on the park are available here.

7.7 Side road to exhibits on the rim of the Colorado River Gorge.

8.0 Cross The Neck, the narrow single connection of the promontory ahead with the Big Flat area. Drainage to the west is into the Green River and that to the east is into the Colorado River (fig. 6.18). An overlook area on the east, at the southern end of The Neck provides a view of switchbacks in the Shafer Trail and the lower part of the trail on the White Rim. Beyond the view point the park road climbs through double cuts of Navajo Sandstone.

9.2 The park road passes between outliers of the upper part of the cross- bedded Navajo Sandstone.

12.3 Beginning of twisting road through outcrops of Navajo Sandstone which are overlain, in part, by recent deposits of windblown sand.

13.6 Mesa Trail parking area. The trail leads a short distance to the east to a viewpoint over the canyon rim.

13.9 Junction of Side Road West to Upheaval Dome and Green River Overlook with the Main Park Road. Turn westward toward Upheaval Dome (fig. 6.19). Navajo Sandstone outliers toward the north are capped by flat-lying beds in the lower part of the formation.

14.2 Junction of roads to Upheaval Dome and Green River Overlook. Turn to right (northwest) toward Upheaval Dome.

15.4 The road now drops into a minor tributary canyon eroded into well-bedded Kayenta rocks below the cross-bedded Navajo Sandstone.

Figure 6.18. View northeastward from the lookout near the Island in the Sky Ranger Station, at Mile 7.7, into the gorge of the Colorado River. Shafer Trail is the road on top of the White Rim Sandstone in the distance. Rocks ranging from the Culter Group up through Wingate Sandstone are exposed along the walls of the gorge. The La Sal Mountains are in the background.

Figure 6.19. Low-angle aerial photograph of Upheaval Dome and the canyon of the entrenched Green River beyond. Wingate Sandstone forms the inner cliffed rim of Upheaval Dome and is surrounded by a prominent ridge of Navajo Sandstone in the rim syncline of the uplift. Wingate Sandstone forms most of the angular cliffs along the gorge of the Green River in the background and toward the left (photograph by W.K. Hamblin).

16.9 Overlook into Green River to the southwest. White Rim Sandstone forms the lower terrace above the inner gorge. Moenkopi and Chinle Formations form the slope above the White Rim and below the vertical Wingate Cliff. Entrenched meanders along the Green River are outlined by the White Rim outcrop pattern.

17.6 Rocks along the road now begin to dip moderately steeply toward the northwest into the syncline which rings around Upheaval Dome.

17.9 Cross onto Lower Navajo Formation which rings the dome like a doughnut caught in the surrounding syncline.

18.4 Parking area for Whale Rock Foot Trail. Whale Rock is an erosional remnant of Navajo Sandstone about on the axis of the rim syncline. Sheer cliffs of Wingate Sandstone are visible in the Green River drainage to the south.

18.9 The road crosses from Navajo Sandstone onto Kayenta Formation which here forms a limited subsequent valley around the uplifted core of the dome, below Navajo Sandstone and above Wingate Sandstone (fig. 6.19).

19.2 End of Road and Picnic Area at Upheaval Dome. Crater View Foot Trail leads approximately 500 yeards toward the northwest to the rim of the core of Upheaval Dome. Steep southeast dips of the Kayenta Formation show near the parking area and along the trail.

The crater rim is formed of resistant Wingate Formation. Light-colored Chinle Formation and reddish Moenkopi beds form much of the sharply dissected cone in the interior of the dome. Permian Cutler rocks are the oldest ones exposed in the dome and are exposed only in the innermost part of the uplift, where deep canyons have cut across. For some years the dome was considered to be a salt dome but opinion now favors a meteorite impact origin for the peculiar circular uplift.

Return to the junction with the Green River Overlook road.

24.3 Junction, Turn West toward Green River Overlook. The road is at about the contact of the Navajo Sandstone on the Kayenta Formation.

24.8 Descend over a cross-bedded sandstone ledge, in the transition at the Kayenta-Navajo contact, onto Kayenta beds which form the flats to the west.

25.2 Bend in the road at the west edge of the flat. The campground area is a short distance to the south at the east edge of a Navajo Sandstone bluff. The road to the viewpoint descends from the campground terrace onto Kayenta beds and continues to the south.

25.7 End of Road at Green River Overlook. White Rim Sandstone forms the prominent rim of the inner gorge in the valley below (fig. 6.20). Wingate Sandstone forms the vertical cliff below the parking

Figure 6.20. View southeastward from the Green River Overlook at Mile 25.7. White Rim Sandstone forms the rim of the inner gorge of the canyon, beneath slopes on the Moenkopi and Chinle Formations. Wingate Sandstone forms prominent outliers and cliffs in the background as well as the rocks in the immediate foreground.

area and above the slopes of Chinle and Moenkopi Formation. The Moenkopi beds here are tan and grayish green unlike most outcrops of the formation in the Colorado Plateau.

Turn around and return to the main park road.

27.5 Junction of Upheaval Dome-Green River Overlook Road with the Main Park Road. Turn south toward Grand View Point. Cross-bedded Navajo Sandstone is exposed on either side of the road south of the intersection. The lower part of the formation is bedded in relatively uniform cross sets, but the upper part has more eolian crossbeds for which the formation is famous.

30.1 Murphy Point Overlook Trail Junction with the main park road. The side road is a four-wheel-drive quality road.

31.7 Side road junction to picnic area. Continue straight ahead toward Grand View Point. The road is on Kayenta Formation.

33.5 Overlook to the west into the Green River drainage.

33.9 End of the Road at Grand View Point. Junction Butte is the outlier of Wingate Sandstone to the south. The White Rim separates the upper and lower parts of the canyon along both the Green River, to the west, and the Colorado River, to the south and east. Reddish Cutler rocks below the White Rim form some of the spectacular country to the south (fig. 6.21). Cutler redbeds inter-finger with the light-colored Cedar Mesa Sandstone in the area beyond the river, to the south, in the Needle section of the park and produce some of the candy-striped country visible from here. The Abajo Mountains are on the skyline beyond the Needles area. Turn around and return to Utah State Highway 313 and U.S. Highway 163.

Figure 6.21. Southward from Grand View Point toward the Abajo Mountains, over the gorge of the Colorado River. White Rim Sandstone caps the plateau surface in the foreground and intermediate distance. Banded outcrops in the far distance in front of the Abajo Mountains are in the Cedar Mesa Sandstone in the Needles section of the national park.

Segment 6B

0.0 Road Junction Arches Monument Headquarters. Turn North (Right) Off U.S. Highway 160 into the headquarters area. Rounded rocks immediately behind the residence areas are in the Navajo Sandstone. Stop and pay one dollar fee at the entrance booth near the Visitors Center.

0.4 Visitors Center. View of the Moab Valley Fault with Entrada Sandstone faulted against Hermosa-Cutler-Rico Formations to the south.

0.7 Base of the Carmel formation and Top of the Navajo Sandstone. Navajo beds are exposed in vertical cuts on both sides of the road. The crinkly character of the Carmel beds shows well on the skyline.

1.6 Pull Out at Lookout Point 2. To the west, back down the road, switchbacks can be seen carved in the Navajo Sandstone and the basal beds of the crinkly Carmel Formation from the Visitors Center to the southwest to here. Massive Entrada Sandstone forms the cliffs to the north and west of the road. White cross-bedded Navajo Sandstone forms exposures below. The nonresistant crinkly beds which weather into reddish slopes are the Carmel Formation. The abrupt termination of the Entrada Sandstone against the Rico-Hermosa sequence can be seen just at the narrows in Moab Canyon beyond the Visitors Center to the southwest or just south of the first bend of the National Monument road beyond the Visitors Center. This expression is of the West Spanish Valley or Moab Valley Fault which here has dropped the Entrada Sandstone and underlying rocks down 2 to 3 thousand feet with reference to those formations exposed to the south of the fault.

Brick red, rounded arkosic sandstone of Cutler Formation is seen beyond the narrows in Moab Canyon where the Hermosa rocks are exposed just south of where U.S. Highway 163 disappears up Moab Canyon. Cutler rocks are overlain at about midslope by a thin sequence of Moenkopi Formation, and it in turn by the gray slope and ledges of the Chinle Formation. Angular jointed Wingate Sandstone forms the vertical cliff and skyline but is capped here and there by thin, flaggy remnants of Kayenta Sandstone. Recent sand dunes can be seen between the highway and railroad track to the south and just beyond the railroad tracks in the vicinity of the trace of the Moab Valley Fault. Ragged outcrops between the highway and the railroad to the south are Entrada beds badly broken along the Moab Valley Fault. Trace of the fault is approximately along the railroad track.

A small fault can be seen to the northwest of this stop, where the Carmel Formation abruptly terminates against massive beds of the Entrada Sandstone. Displacement of this small fault here is approximately 50 or 60 feet. This small fault is well exposed at the east end of the road cut east of this stop. Slickensides, grooved shiny surfaces, are exposed in the cut and along the fault to the east and west.

Outcrops of Carmel Formation can be seen to the north beneath the Entrada behind the Navajo Sandstone beds exposed in the immediate road cut. Sandstone Peaks of the La Sal Mountains can be seen to the east through the road cut in the Navajo Sandstone. **Continue ahead into the National Monument Area.** The east end of the road cut is in the Carmel Formation which is now exposed as red "stone babies" to the north of the road. The road continues in the uppermost beds of the Navajo Sandstone patterned into characteristically rounded, eroded remnants.

2.6 Geologic Stop. Pull to the North (Left) into the Parking Area. Turnout 4. Entrance to the south end of Wall Street (fig. 6.22). Walk a short distance to the overlook into the joint-controlled, fin-bounded canyon. Early visitors imagined the skyline resembled that of a city. Return to the cars and turn left (east) onto main Monument road.

3.7 View east down the road toward the La Sal Mountains. Excellent exposures of Navajo Sandstone show the broad synclinal character of the Courthouse Wash structure.

4.2 Cul De Sac Turn Around of the La Sal Mountain Overlook. Courthouse Towers can be seen to the west (fig. 6.23), including the Three Gossips (fig. 6.24), Tower of Babel, and Organ Rock. To the east, this view shows the characteristic broad downbuckled syncline development of the Navajo Sandstone. Turn around and return to the main Monument highway. The main park road continues through road cuts in the Carmel Formation and uppermost beds of the Navajo Sandstone and descends down the north-

Figure 6.22. Northward along Wall Street from the parking area. Massive sandstone cliffs are in the Moab Tongue of the Entrada Formation above the crinkly, moderately well-bedded, Dewey Bridge Member of the formation.

Figure 6.23. Aerial view southeastward to the park entrance area and Moab Valley from the northwest. The highway into the interior of Arches National Park winds through the buttes and mesas in the middle distance. Colorado River leaves Moab or Spanish Valley through the V-shaped notch in the center distance. Prominent joints in the foreground are in the upper part of the Moab Tongue of the Entrada Formation.

dipping top of the Navajo Sandstone into Courthouse Wash Syncline.

Figure 6.24. The Three Gossips as seen from the park road at approximately Mile 4.5. These monoliths are in the Moab Tongue of the Entrada Formation.

6.4 Cross Courthouse Wash. The wash is entrenched here into the Navajo Sandstone as well as in a thick, vertical-walled alluvial cover.

7.1 The road passes near a small isolated pinnacle of crinkly-bedded Carmel Formation. To the east the hummocky "beehive" surface of the Navajo Sandstone shows well rising on the north flank of the Courthouse Wash Syncline. The road continues ahead on the top of the Navajo Sandstone.

7.9 Turnout Point 11. Petrified dunes viewpoint. Cross-bedded Jurassic Navajo Sandstone shows in the ancient exhumed dune fields to the east. Jointed Entrada Sandstone can be seen to the northeast in the windows section of the park.

9.5 Massive white sandstones of the basal Entrada Formation in the pinnacle area appear to have been channel sands and maybe have differentially loaded the soft underlying sediments and produced some of the crinkly character of the Entrada. At least two irregular deep penetrations of the channel sandstones down into the lower beds can be seen. Continue ahead with outcrops of basal beds of the Entrada Formation on both sides.

10.5 Turn Out at Balanced Rock (fig. 6.25). Ham Rock is perched high on the Entrada Sandstone outlier the east. The road continues ahead on the stripped surface of Navajo Sandstone or in basal beds of the Entrada Formation. Designations of the Entrada, Carmel, and Navajo Formations are shown at the east end of the pull out.

Figure 6.25. Ham Rock, a balanced-rock outlier of the lower part of the Moab Sandstone Tongue, is perched on crinkly Dewey Bridge beds of the Entrada Formation as seen from Mile 10.5.

11.2 Junction of Road to Double Arch and Windows Section of the Park to the East. Turn Right To the northeast, just beyond the road junction can be seen the south flank of Salt Valley Anticline with Wingate Sandstone beds dipping to the southeast. The road continues to the east of the junction through cross-bedded Navajo Sandstone.

11.5 Turnout Point 22. Strongly-jointed Entrada Sandstone can be seen to the north beyond Salt Valley, beneath the white cap of less intensely-jointed Curtis Sandstone.

12.4 Garden of Eden Viewpoint to the east. "Elephants" are in the Moab and Dewey Bridge Members. Cross-bedded Navajo Sandstone forms the pedestals at their base.

13.5 Beginning of one-way road. Turn around at the Double Arch and Window Arches. Turnout 18. Trails to Turret Arch and North and South Window Arches to the east. A maintained trail leads southeast from the parking lot into the arch area. One can see Turret Arch to the southwest from the trail. A small keyhole arch occurs in basal beds of the Entrada Sandstone. South Window Arch is a high round arch in basal beds of the Moab Sandstone with the Dewey Bridge beds depressed below it in a rather characteristic crinkly pattern. Some of the basal beds of the Entrada Formation may be part of the crinkled lower beds of the shear, cliff-forming Entrada. A short trail leads from the main trail to North Window Arch. An excellent view of the valley to the east can be seen through North Window Arch.

14.0 Parking zone, Turnout Point 19. A trail leads past alcoves to Double Arches (fig. 6.26) to the northwest of the parking lot. **Return to the Junction with the Main Road Near Balanced Rock at Mile 11.2.**

11.2 Junction of the Windows Section Road with the Main Monument Road into the Devil's Garden and Delicate Arch Sections. Turn Right (north) through cuts in the Navajo and Kayenta Formations.

12.5 Panorama Viewpoint, turn off to east (right). To the northwest we can see the downfaulted valley block with Jurassic and Cretaceous rocks dropped within the Salt Valley graben. The high pinnacled fin development is in the Entrada Formation. The Entrada beds are overlain by the white Curtis Sandstone and it is overlain by various colored shales of the Morrison Formation. Some gray and yellow brown weathering Mancos Shale can be seen in the valley to the north.

Figure 6.26. Double Arches as seen from the southeast from near the end of the trail that leads from the parking area at Mile 13.5. The arches, like many other features in the park, are in massive Moab Sandstone Tongue of the Entrada Formation.

13.9 Gravel turn out to the east. Gray and yellow gray hills down the valley to the east beyond the dune-covered brush area to the north are in the Mancos Shale, north of the main Salt Valley Fault. To the east the prominent high cliff is in the Wingate Sandstone which is underlain by the reddish slope of the Chinle Formation. Both the Wingate and Chinle exposures are south of the Salt Valley Fault.

14.5 Dakota Sandstone and Morrison Formation exposed to the north of the road, across the gully. These formations are the ashy-appearing beds south of the prominent sandstone in the core of the Salt Valley Anticline. Prominent white Curtis Sandstone cliffs to the north cap the top of the pink massive Entrada Sandstone, the sandstone in which most of the arches and the fins of the park have been carved.

15.8 Turn out on the north (right). View of pinnacles of Entrada Sandstone beneath the massive, light-colored, Curtis Sandstone. Colored Morrison, and gray Dakota and Mancos beds are exposed in the foreground and in the gully. Mancos beds are exposed in the bottom of the gully. Entrada beds form the castellate surfaces along the skyline.

16.5 Salt Valley Overlook, Turn to the Right (East) on to Short Road. Beds near the anticlinal structure show very well, with formations from the Entrada Sandstone to the Mancos Shale dipping steeply to the south, off the south flank of Salt Valley Anticline. A fault, visible east of the parking area, repeats the white Curtis Sandstone. Salt Valley to the southeast is carved in the Morrison and Mancos Formations. To the south the low escarpment on the south side of Salt Valley is in Chinle Shale and Wingate Sandstone, but the latter does not form the characteristic bold escarpment here, like it does to the east apparently because of faulting and breaking of the beds.

16.8 Turn Off to the Right to the Fiery Furnace Parking Area. The La Sal Mountains rise to the east above the folded Mesozoic rocks in the Salt Valley Anticline structures. Ahead and to the north the castellate Entrada forms the fins and stone babies in the immediate vicinity. The finned Fiery Furnace area is relatively complex topographically. It is a primitive area and the possibility of becoming lost is great. As a consequence only ranger-conducted tours are led into the Fiery Furnace area. Fins are in parallel jointed Moab Sandstone of the Entrada Formation.

17.1 View to the south shows the dip of the faulted sequence of Carmel, Entrada, Morrison, Summerville, and Mancos Formations all down-faulted against the Chinle beds which form the escarpment to the south. Navajo Sandstone is exposed in the gullies immediately south of the road. The Dewey Bridge forms the crinkly-bedded zone at the base of the Entrada cliff to the north.

17.7 Upper Fiery Furnace Turn Out. The thin finlike character of the resistant Entrada Sandstone shows well above the crinkly Dewey Bridge beds. Road cuts to the west are in the Dewey Bridge with Navajo Sandstone exposed in gullies to the south.

18.8 Viewpoint and parking area to the north (right). Maintained trails lead to Sand Dune Arch and Broken Arch to the north. Broken Arch can be seen directly to the north, at the end of a fin of Entrada Sandstone, beyond the grassy sagebrush-covered flat. Sand Dune Arch is in the same general vicinity.

19.5 Skyline Arch Turn Out. The road continues in upper beds of the Navajo Sandstone with crinkly Dewey Bridge beds exposed just above, beneath the walled cliffs of the Entrada Formation.

20.1 Beginning of One-Way Loop at the Entrance to Devil's Garden. Fins of Entrada Sandstone are well developed in areas to the east. **Walk-in Picnic Area.** Junipers and pinion form shady nooks in the sand-covered valleys between fins of Entrada Sandstone.

20.2 Junction of Road to Devil's Garden Campground. Turn right for campground, rest room and drinking water facilities.

20.8 Parking Area at the Entrance to the Devil's Garden Trail to the west through Fin Canyon. Fins in the vicinity of the parking area are all in the Entrada Formation (fig. 6.27). Trails lead to Landscape Arch (fig. 6.29) and other arches to the west. Return on the main road toward the headquarters.

26.9 Junction Road to Delicate Arch.

Figure 6.27. Joint separated fins of massive sandstone of the Entrada Formation as seen westward towards the Fin Canyon Trail area in the Devil's Garden. The Fin Canyon Trail leads through the pass between the fins to the left of the center in the photograph.

Figure 6.28. Landscape Arch is the longest and highest arch in the national park. The arch is reached by a trail westward from the Devil's Garden area through Fin Canyon.

Turn Left. Gray Morrison beds are exposed across the canyon to the north. The road cuts through the lower part of the Mancos Shale and descends to Salt Valley Wash, primarily in the Mancos Shale. Upper beds of the Dakota Sandstone and the Morrison Formation are exposed to the north.

27.4 View to the southeast of the Wingate Cliff above the Chinle and Moenkopi slopes, south of the Salt Valley Fault. The road continues in the Mancos Shale. A brown-weathering sandstone in the Dakota Sandstone forms the resistant cuesta to the north and northeast.

28.0 Cross Salt Valley Wash.

28.2 Delicate Arch Trail Parking Area to North. Turnbow Cabin, one of the early pioneer cabins, can be seen to the north, by the old pole corral. The steep southward dip here in the maroon shales and white sandstones of the Morrison Formation is typical of the south side of the Salt Valley Anticline. The pink "slick rock" rim to the north is formed of southward dipping Entrada Sandstone. Prominent high pinnacles to the east are in the Entrada Sandstone and surround Delicate Arch which is the narrow fin about in the middle of the promontories in the skyline.

28.3 Cross Culverts on Salt Wash.

28.6 Road is on gray Mancos Shale, tan Dakota Sandstone forms the first prominent hogback to the north (left) above the white and brown ledgey-appearing beds of the Morrison and Cedar Mountain Formations which form the high cuestas on the skyline to the north. To the south, across Salt Valley, Mancos Shale is faulted against basal beds of the Chinle and Moenkopi Formations. The red rocks occur on the south side of the Salt Valley Fault.

28.9 Cross the Dakota hogback. The road continues on basal beds of the Mancos Shale, along the base of the striking hogback in the Dakota Sandstone.

29.1 Cross back through the Dakota hogback onto soft gray green shales of the Cedar Mountain Formation and then through a white sandstone in the Morrison Formation. This sandstone forms the crest of the high hogback to the west. The road continues in the Morrison Formation through the narrow part of the canyon and then swings back to the southwest in a subsequent valley in the Morrison Formation.

29.3 Delicate Arch Viewpoint. Delicate Arch is in the Entrada Sandstone in the pinacles to the north. Ragged cherty beds of the Morrison are exposed immediately north of the parking area with "slick rock" of the Entrada Formation exposed below it and on to the north. Summerville beds are the deep brick red beds on top of the slick rock zone, beneath the cherty units of the basal conglomerate of the Morrison Formation. To the east, beyond the parking area, the down faulted graben shows very well, with Wingate cliffs on the north and Chinle and Wingate cliffs on the south side, with downdropped Jurassic rocks in the center of the structure. Tilted beds in the immediate foreground are in the Morrison Formation. Return to Visitors Center and to U.S. Highway 163.

Segment 6C

Potash-Shafer Dome

0.0 Junction of Utah State Highway 279 and U.S. Highway 163. Utah Highway 279 leads down the west side of the Colorado River to the mining operations at Potash. To the east can be seen the La Sal Mountains on the skyline in essentially three clusters as North Mountain, Middle Mountain, and South Mountain, beyond the Triassic cliffs of the north side of the Spanish Valley Anticline. To the south the railroad is in the Honaker Trail-Cutler sequence which is overlain by Chinle, Wingate and Kayenta Formations. Moenkopi beds are cut out here by an angular conformity between the Cutler and Chinle Formations. The railroad was put in to serve the potash mines at Potash, sixteen miles downriver to the south.

0.6 Exposures of laminated white and red beds in the upper part of the Cutler sequence on the southwest side of Moab Valley near Utah Highway 279, to the west. Tailing pits of Atlas Uranium Ore Reduction Plant continue to the right. Faulted Chinle and Wingate beds are exposed in the vicinity of the railroad.

1.0 Colorado River swings in close against the road. Point bar development is obvious on the eastern inside of the meander bend. Directly downstream, the Portal of the exit of the Colorado River from Moab Valley is capped by angular-weathering Wingate Sandstone, with Chinle beds forming the underlying slope. Navajo Sandstone is exposed on the skyline downstream a short distance.

2.0 The light-colored beds immediately east of the river by the telephone poles are in the Paradox Formation. These rocks have pierced upward along the Moab Valley or Spanish Valley Fault System and are exposed several thousand feet above their normal position. No such piercement type features are visible on the western side of the river.

2.7 Base of the massive Wingate Sandstone on variegated Chinle Shales. The Wingate Sandstone is more prominently bedded here than it is in many areas, perhaps because this is its expression in the narrow canyon.

3.1 Indian ruins (probably granary structures) high on the east side of the canyon are beneath the overhanging ledge of the Kayenta Sandstone above the Wingate cliffs. Wingate Sandstone forms a massive cliff along both sides of the river and is here blanketed by desert varnish. Ahead hummocky round hills of Navajo Sandstone occur on top of the flaggy-bedded Kayenta Formation.

3.5 Alcoves and hanging gardens can be

Figure 6.29. Navajo Sandstone exposed along the gorge of the Colorado River below Moab. Gravel-capped terraces have been etched across the massive Navajo Sandstone as exposed here in the King's Bottom Syncline.

seen in the Navajo Sandstone, the formation that forms the rounded vertical cliffs on both sides of the canyon (fig. 6.29). Terraces blanketed with gravel continue downstream on both sides of the canyon at varying elevations up to 50 feet above the river.

4.3 Massive Navajo Sandstone. Some erosional arch features show well in the sandstone. Fin development shows in the Navajo Sandstone on the east wall of the gorge along the strong joint system which is parallel to the Spanish Valley Fault to the north. Navajo Sandstone usually shows jointing more spectacularly than other units in the Mesozoic sequence.

4.8 Indian petroglyphs chipped in desert varnish in the Navajo Sandstone 10 to 15 feet above the road on the right. These petroglyphs are typical Fremont carvings. Fins of the Navajo Sandstone are well developed along the side of the gorge beyond Kings Bottom. The rocks are nearly flat lying here in the trough of the Kings Bottom Syncline.

5.5 Kane Springs Creek enters the Colorado River from the east. Dinosaur tracks and Indian petroglyphs are visible in Kayenta beds and blocks to the north. Complexly cross-bedded and in some cases contorted-bedded and burrowed, tracked and trailed Kayenta Sandstone is exposed at river level and near the turn out. High alcoves and arches developed in the Navajo Sandstone can be seen high on the canyon wall to the east. The Kayenta Formation is expressed here in the typically ledgey slope zone which is capped with much high-level terrace gravel marking former positions of the Colorado River.

6.6 The canyon rim to the south is formed in the slabby Kayenta Sandstone. Because of the northward dip in the meander around Amasa Back, the long meander core of Kayenta Sandstone is at river level, and is alternatingly high above river level and near river level again as the channel swings back and forth from south to north then south again.

8.8 Base of the Navajo Sandstone is exposed at road level. For the next half mile Navajo Sandstone forms the outcrops in the road cuts on the north side. Some Kayenta beds are still exposed above river level on the southwest bank, with massive "beehives" of Navajo Sandstone developed high on the skyline ahead and to the south.

9.8 Bootlegger Canyon enters the canyon from the north. A road leads up the canyon to near Little Rainbow Bridge.

10.1 The river now swings approximately parallel to the branch line of the D&RGW Railroad leading from Crescent Junction south to Potash. The railroad tunnels through the long spur of the meander of the river. The railroad tunnel is approximately 1 mile long and opens into the headwaters of Bootlegger Canyon. Kayenta rocks are exposed both east and west of the river with gravel terraces well developed to the east on the point of Amasa Bank.

11.4 Very narrow Day Canyon leads off to the west (right). Parallel jointing is well expressed in the vertical walls of Wingate Sandstone.

12.8 Channel sandstone filling in uppermost beds of the Chinle beneath the massive shear wall of Wingate Sandstone exposed in railroad cuts to the southwest (right). To the left the massive wall of Wingate Sandstone is interrupted by a semiledge zone of well-bedded Kayenta rocks which are overlain by massive Navajo Sandstone near the top. The wall is called The Billboard.

13.0 Jug Handle Arch Visible High and to the Right in the Wingate Sandstone.

13.4 Major Tributary of Long Canyon Enters from the West. A primitive road from Big Flat and Dead Horse Point comes down Long Canyon on top of the gorge to the west. Chinle beds are exposed at the Long Canyon Junction and are also well exposed on the east side of the canyon. Beds dip sharply northward off the Kane Springs Anticline into the Kings Bottom Syncline.

14.3 Storage bins and processing plant of Potash development are directly ahead. Moenkopi rocks should appear near here but are not exposed through the thick cover.

14.5 Top of the Cutler beds is exposed at road level at the north end of the Potash railroad yards. The overlying Moenkopi Formation is thin here and forms the brownish shaly slope above the brick red units of the upper part of the Cutler sequence.

14.8 Massive red sandstones in the Rico Formation are exposed in low exposures away from the river. The high rim at the margin of Big Flat is in Wingate Sandstone on the skyline to the west.

15.3 Potash Plant to the Right. End of paved road. Return to U.S. Highway 163.

Segment 6D

0.0 Junction of State Highway 211, with U.S. Highway 163. Continue westward on State Highway 211 in the western part of Church Rock Valley. The highway is constructed within the Carmel Formation at the base of the Entrada Sandstone which forms the rounded "beehive" outliers. Entrada Sandstone forms the rim to the west, as well as the massive light bluff to the east beneath the wooded slope of the Morrison Formation.

0.6 Buildings to the south are part of Ogden Center which was established by a Mrs. Ogden who came west from Pennsylvania and started a religious group here. The sect more or less disbanded following her death.

3.3 Summit. Buildings here are also part of Ogden Center. Light-colored, cross-bedded Entrada Sandstone forms the lower bluff on the north and is overlain by a thin reddish Summerville Formation. Morrison Formation caps the hills.

5.5 Navajo Sandstone is exposed in gullies to the northwest. The highway is constructed in a subsequent valley carved in the less resistant Carmel Formation. Bluffs to the south expose the light-colored, cross-bedded Entrada Sandstone at the base and a wooded slope above carved, in large part, on Morrison beds. Dakota Sandstone caps the rim. The Abajo Mountains, toward the west and southwest, are cored by Tertiary igneous intrusions.

7.6 Bend in the road at approximately the drainage crossing is about on the contact of Carmel redbeds on light-colored Navajo Sandstone. The road climbs onto the dip slope of the Navajo Sandstone to the west.

10.3 Summit. The road now descends quickly into Indian Canyon which is cut into Navajo Sandstone here.

11.5 Cross a small drainage. North of the road upper beds of Navajo Sandstone are well exposed in the bluff. South of the road the top of the Wingate Sandstone and base of the Kayenta Formations are about at road level, on an upfaulted block. The fault trace is at about the break in slope on the south, ahead, and continues westward toward the mountains.

11.9 Cross a large culvert in a sharp switchback on the road. The fault trace is immediately south of the road fill, with Navajo Sandstone on the north faulted down against Wingate and Kayenta beds on the south. The road swings north of the fault beyond the bend and drops down into Kayenta beds.

12.8 Overhang and cattleguard. Cliffs near the road are in Wingate Sandstone (fig. 6.30). The high Abajo Mountains to the south are the source of most of the creeks of the region.

Figure 6.30. Rounded massive Navajo Sandstone caps the sheer canyon wall west of the petroglyphs in Newspaper Rock State Park at Mile 13.0. Wingate Sandstone forms the prominent massive lower cliff and is separated from the upper Navajo beds by moderately well-bedded Kayenta Formation.

13.0 Turnout to Newspaper Rock in the state park. Newspaper Rock is a group of petroglyphs carved through a veneer of desert varnish on jointed Wingate Sandstone (fig. 6.31). Some of the older inscriptions are of Fremont culture, particularly the ones with triangular bodies. Mounted figures may represent Ute Indains and are probably 200 years old.

15.0 Reddish rocks near road level and in the tributary canyons to the south are upper beds of Chinle Formation.

17.3 Angular jointed Wingate Sandstone forms the prominent cliffs on either side of Indian Canyon. Large blocks of Wingate Sandstone topple when the softer, more easily eroded, Chinle Formation below is removed. As a consequence blocks and boulders of sandstone generally litter the Chinle slope. Slabby slopes above the Wingate Cliff are in the Kayenta Formation.

20.1 End of State Highway 211, beginning of park access road. North and South Sixshooter Peaks are the prominent outliers of Wingate Sandstone to the north.

20.6 Side road to Dugout Ranch, almost at the east end of a small reservoir on the northeast.

Figure 6.31. Newspaper Rock in the state park. Petroglyphs have been chipped through the brown veneer of desert varnish on the Wingate Sandstone and show mixed Fremont and Ute Indian cultures.

21.1 Dam on the reservoir. Shinarump Sandstone now appears above the valley fill of Indian Creek toward the west. Pink and maroon varicolored Chinle beds are exposed in bluffs beneath jointed Wingate Sandstone along the margin of the valley (fig. 6.32).

22.0 Shinarump Sandstone is exposed in road cuts and in the gullies on either side of the road as well as in major exposures along Indian Creek to the west.

23.0 The road crosses through Shinarump Sandstone and onto Moenkopi beds. Shinarump beds form the low, light-colored, cross-bedded rim across Indian Creek and

Figure 6.32. Massive Wingate SAndstone cliff rises above slopes on the Chinle Formation near Dugout Ranch at approximately Mile 21.

near the road, and the unit thickens toward the northwest from here.

25.0 North and South Sixshooter Peaks, to the west, are outliers of Wingate Sandstone above slopes on Chinle Formation. Shinarump Sandstone forms the prominent bench at the base of the light-colored Chinle and above the more distinctly-bedded Moenkopi rocks. Moenkopi redbeds are exposed as alternating low ledges and slopes near the road and to the east.

27.1 Road cuts through upper Organ Rock redbeds consisting of alternating orange red sandstone and more maroon to purplish red arkosic sandstone.

27.8 Cross Indian Creek which here is eroding into Organ Rock redbeds. The highway swings to the northwest beyond the bridge and parallel to exposures of the Cutler Group.

28.6 Well-exposed Organ Rock redbeds in bluffs on the west. A tongue of Cedar Mesa Sandstone is exposed near valley level toward the north.

30.0 Cedar Mesa Sandstone is exposed at the base of North Sixshooter Peak to the south of the road (fig. 6.33), and is here muddy and mottled with redbeds. Moenkopi redbeds, above the reddish Cutler sequence, form thin ledges and slopes beneath the shelf held up by Shinarump Sandstone. Individual lenses and tongues of Cedar Mesa Sandstone thicken toward the west and exposures here appear to be at the boundary between the eastern redbed facies and the western light-colored, possibly eolian, sandstone facies.

Figure 6.33. North Sixshooter Peak, an outlier of Wingate Sandstone, as seen from the northeast at approximately Mile 30. The lower few ledges are on the reddish Cutler beds, of Permian age, and are overlain by the prominent, well-bedded, relatively light-appearing Moenkopi Formation. Shinarump Sandstone forms the prominent ledge in the middle part of the slope and is overlain by light-colored gray slopes of the upper part of the Chinle Formation. Uppermost Cedar Mountain Formation is exposed at the base and produces the flat country in the foreground.

30.8 Cedar Mesa Sandstone is exposed at road level and in the first ledges above the road on the south. Organ Rock redbeds occur above in rhythmic alternation of orange red and purplish red sandstone and siltstone. Toward the north Junction Butte and Grandview Point are the promontories on the west side of the Colorado River Gorge and are the termination of the long continuous Wingate Sandstone cliff. They are in the Island in the Sky Section of the park. The road continues toward the west through well-bedded Cedar Mesa Sandstone.

31.4 Cross a drainage cut through rounded exposures of Cedar Mesa Sandstone. A short distance to the east these sandstone beds wedge out into red Organ Rock beds. Toward the west, however, Cedar Mesa Sandstone thickens and becomes light tan to gray and is highly cross-bedded, in general with an appearance like many outcrops of Navajo Sandstone.

33.7 Cross cattle guard, still in Cedar Mesa Sandstone.

34.3 East Boundary of Canyonlands National Park. The Needles Section of the park is now visible to the west with pinnacles and spires in Cedar Mesa Sandstone.

35.2 Side Road to the North to Canyonlands Resort, situated beneath bluffs of Cedar Mesa Sandstone. Gasoline, groceries, rooms, airplane charters, etc., are available in this service area.

35.7 Cross Salt Creek drainage. The park road continues through spectacularly cross-bedded Cedar Mesa Sandstone.

36.5 Side road to the north leads to Canyon Overlook near where Salt Creek plunges over the resistant rim of the inner gorge to the Colorado River.

37.2 Junction of Paved Side Road 0.7 Mile to the South to Cave Spring Ranger Station and Exhibits. Continue ahead on the main park road through cuts in Cedar Mesa Sandstone.

38.6 Wooden Shoe Arch Overlook. Wooden Shoe Arch (fig. 6.34), to the south, is in upper Cedar Mesa Sandstone as are all the exposures near here.

39.0 Junction of Paved Side Road which Leads East to Cave Spring Ranger Station.

Figure 6.34. View southward from the parking area to Wooden Shoe Arch. The arch is in one of the prominent light-gray, massive tongues of the Cedar Mesa Sandstone, the banded sandstone which produces much of the scenery in the Needles section of the park.

Continue ahead on the main park road toward the campground areas.

39.3 Junction West to Squaw Flats Campground Area and Access Road to Elephant Hill. Turn west toward the campground area. The road to the north leads to overlooks on the canyon rim in the northern part of the park.

39.7 Junction side road to Campground area A. Continue ahead.

40.0 Junction side road 0.2 mile into Campground area B. The relatively primitive road heads west toward Elephant Canyon and Elephant Hill and the jeep road into the interior of the park. **Continue on the dirt road with caution** through cross-bedded Cedar Mesa Sandstone. The sandstone has eroded to provide overhangs and numerous small reentrants in the camping area (fig. 6.35).

41.3 Winding road on crest of hill. View to the south is into the Needles section (fig. 6.36) and Chesler Park area of the park.

Figure 6.35. Massive, light-colored Cedar Mesa Sandstone in the Squaw Flats campground area at approximately Mile 40.

Toward the north view is down Elephant Canyon. The top of the Elephant Canyon Formation is the gray green limestone in the bottom of the canyon. From here the road winds down into redbeds of the Halgaito Formation below Cedar Mesa Sandstone.

41.4 The top limestone of the Elephant Canyon Formation forms the resistant ledge in the bottom of the canyon west of the road. This is the resistant unit which forms the platform on which much of the scenic area is carved. The Elephant Canyon Limestone is sparsely fossiliferous but in some areas contains crinoid debris, brachiopods, corals, bryozoans, bivalves, and snails. In other areas the top limestone is a relatively unfossiliferous fine-grained burrowed limestone.

41.9 Cross Elephant Canyon drainage on top of the gray limestone. Cedar Mesa Sandstone still forms the light bluffs on the margins of the canyon.

42.8 Picnic and Turn Around Area at the Beginning of the Jeep Trail Over Elephant Hill. Do not attempt to drive on into the interior of the park without a suitable four-wheel drive vehicle. Turn around and return to the Cave Springs Ranger Station access road.

Figure 6.36. View southward up Elephant Canyon into the Needles section of Canyonlands National Park from Mile 41.3. Banded prominently jointed sandstone is in the Cedar Mesa Sandstone. Halgaito redbeds are exposed in the slope in the foreground, above a ledgey limestone which is at the top of the Elephant Canyon Formation.

46.7 Junction off Main Park Road to Cave Spring Ranger Station. Turn off the main park road toward ranger station. North Sixshooter Peak is on the skyline to the east (fig. 6.37) and Wooden Shoe Arch to the south.

47.5 Exhibits and **water** in parking area on the north of the road. Residence area is in Cedar Mesa Sandstone.

48.1 Cave Spring Ranger Station. Turn east onto the dirt road, down the canyon, toward Cave Spring.

48.9 Junction of Jeep Trail south to Angel Arch and other scenic areas up Salt Creek Canyon. The trail is across loose sand and along the valley bottom. Do not attempt to drive up Salt Creek Canyon without a suitable four-wheel drive vehicle.

Figure 6.37. North Six-Shooter Peak as seen eastward over ledges and cliffs of Cedar Mesa Sandstone from Mile 46.7, near the turn-off to Cave Spring Ranger Station.

49.1 Cave Spring Parking Area. Cave Spring is situated under over-hanging Cedar Mesa Sandstone (fig. 6.38), approximately 100 feet west of the parking area. Cave Spring was utilized by ranchers as a camping area before the park was established. A trail leads around some of the interesting erosional remnants of Cedar Mesa Sandstone.

Turn around and return to the ranger station and then to the main park, either to the north or to the west from the station.

Figure 6.38. Eastward from Cave Springs, beneath the overhang, to North Sixshooter Peak. Cave Springs is situated under overhanging ledges of Cedar Mesa Sandstone, where porous sandstone rests on moderately impervious shaly beds which intertongue with the sandstone.

Segment 7

0.0 Bridge Over Interstate 15 at the Hanksville interchange and Junction of Utah State Highway 24 . The interchange is in lacustrine and fluvial Brushy Basin Member of the Jurassic Morrison Formation. Light-colored sandstone lenses are fluvial deposits which interfinger with varicolored mudstone and siltstone. Toward the west the underlying Salt Wash Member of the Morrison Formation forms the irregular dip slope in the foreground. The Reef of the San Rafael Swell (fig. 7.1) forms the flatirons in the background and is composed, in large part, of Navajo Sandstone from this view. Toward the east the persistent ledge-former near the top of the exposures is the Buckhorn Conglomerate, the basal member of the Cretaceous Cedar Mountain Formation (figs. 7.2, 7.3).

0.8 Serpentine ridges toward the west and along the road are held up by sandstones of channel fills in Morrison-age streams. Softer alluvial plain sediments have been eroded away, leaving the channels now expressed in inverted topography with the former low areas standing high.

1.1 Double road cuts through cross-bedded, light-colored Morrison channel-fill sandstone, at about the contact of the upper Brushy Basin with the lower Salt Wash Members of the formation. Ancient paleochannels are well expressed by ridges to the west.

2.7 Double road cuts through light colored Morrison channel fill. Rounded pinkish and purple bluffs of the Brushy Basin beds capped by Buckhorn Conglomerate rise above the highway on the east. The vegetated area on the west is along the San Rafael River. The road to the south is approximately on the top of the Salt Wash Member.

3.8 Road crosses paved segment of old highway in Morrison beds.

4.0 Road now crosses through numerous sandstone lenses in the Salt Wash Member of the Morrison Formation. These show elongate meandering of the paleochannels. Deposits of both braided and meandering streams are evident in the series of cuts.

4.5 Junction of ranch road to the south. Beyond the junction deep cuts through greenish clay and sandstone mark the basal part of the Morrison Formation above the reddish Summerville rocks.

4.8 Contact of the Morrison Formation on gypsiferous Summerville Formation in the lower western end of the deep double road cuts. Tan to reddish brown mudstone is interbedded with knobby and lenticular gypsum in the upper part of the Summerville exposures.

5.0 Bridge Over the San Rafael River. Excellent exposures of laminated Summerville Formation occur below Morrison beds both north and south of the road. Directly ahead the steeply dipping east flank of the

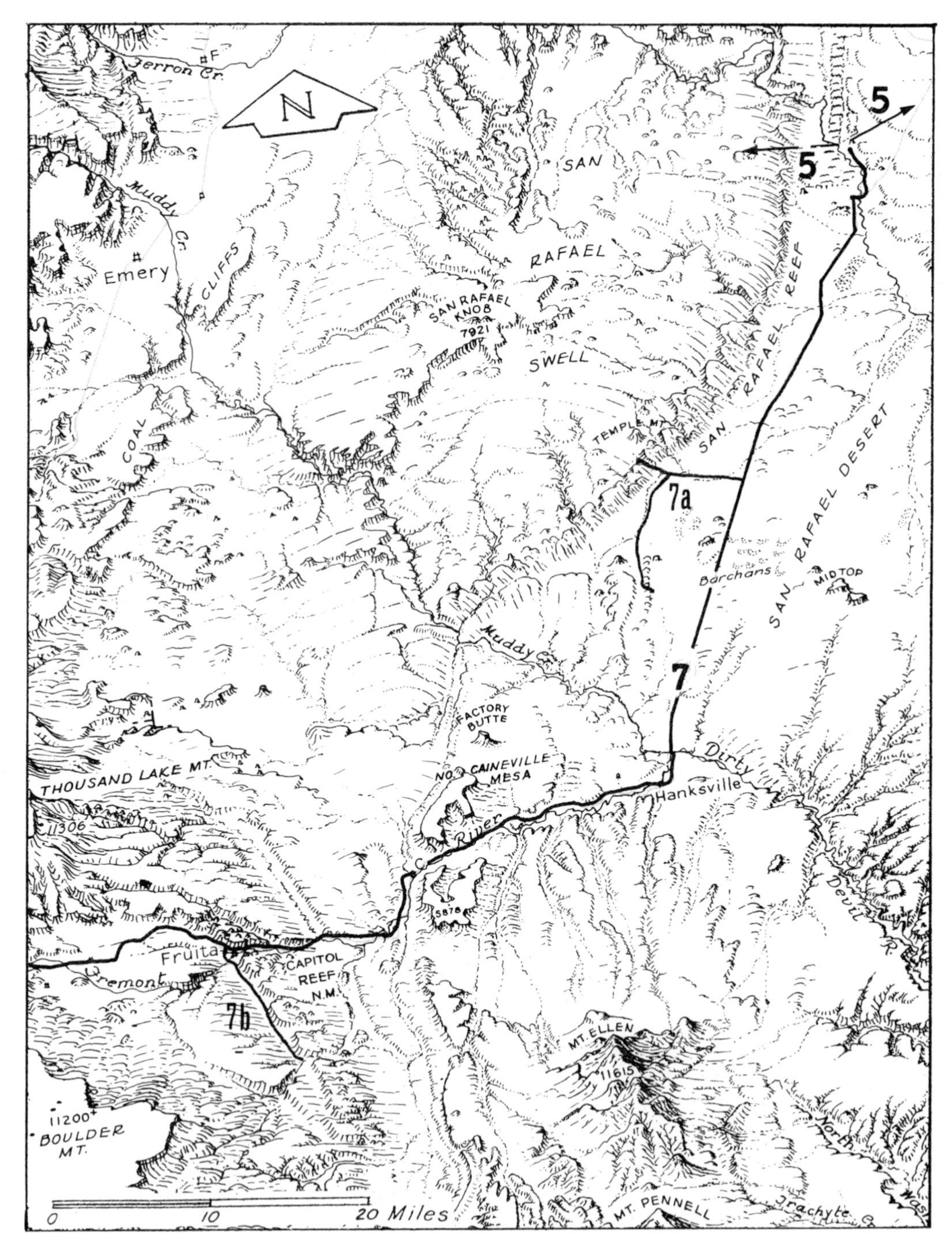

Figure 7.1. Index map to the eastern part of Route 7, from where it leaves Route 5, east of the San Rafael Swell, southwestward to the Capitol Reef area near Fruita. Route 7a leads into Temple Mountain and to Goblin Valley State Park and Route 7b leads along Capitol Reef in the national park.

SAN RAFAEL SWELL

System	Group	Formation	Thickness (ft)	Remarks
PERMIAN		Kaibab Ls	0-85	Park City equivalent
PERMIAN		"Coconino" Ss = undivided White Rim and Cedar Mesa Ss of Cutler	600—800	Diamond Creek Ss of Wasatch Mtns
PERMIAN		Elephant Canyon Fm	500—700	pre-Permian known from limited well data
PENNSYLVANIAN	Hermosa Group	Honaker Trail Fm	200—500	
PENNSYLVANIAN	Hermosa Group	Paradox Fm?	300—500	Oquirrh Fm equivalents
PENNSYLVANIAN	Hermosa Group	Pinkerton Trail Fm	100—200	Manning Cyn equiv
MISSISSIPPIAN		Redwall Ls	600—900	Deseret Ls equivalent; Gardison Ls equivalent
DEV		Ouray Ls	100—200	
DEV		Elbert Fm	200—400	
CAMBRIAN		"Lynch" Dolomite	1000—1300	
CAMBRIAN		"Maxfield" Limestone		
CAMBRIAN		Ophir Fm	200	
CAMBRIAN		Tintic Quartzite	150—300	
P€		"granite"	—	

System	Formation	Member	Thickness (ft)	Remarks
CRETACEOUS	Mancos Shale	Blue Gate Shale M	1500+	
CRETACEOUS	Mancos Shale	Ferron Ss Member	100—300	12' coal near Ferron
CRETACEOUS	Mancos Shale	Tununk Shale Member	400—700	
CRETACEOUS	Dakota Ss		0-50	
CRETACEOUS	Cedar Mtn Fm		0-100	*"gizzard stones"*
JURASSIC	Morrison Fm	Brushy Basin Sh M	250—400	*dinosaur bones*
JURASSIC	Morrison Fm	Salt Wash Ss M	200—300	uranium in Green River mining district
JURASSIC	Summerville Fm		260—330	chocolate siltstones gypsum veinlets
JURASSIC	Curtis Fm		75—250	*Pentacrinus* *Camptonectes*
JURASSIC	Entrada Ss		300—800	poorly cemented earth- source of modern dune sands of south San Rafael area
JURASSIC	Carmel Fm (=Twin Creek, lower Arapien)		150—650	gypsum; red and green shale; *Trigonia* *Ostrea* *Camptonectes*
JURASSIC – TRIASSIC	Navajo Ss		440—540	highly crossbedded
TRIASSIC	Kayenta Fm		40—240	these sandstones form the reef along the southeastern margin of the San Rafael swell
TRIASSIC	Wingate Ss		350-400	
TRIASSIC	Chinle Fm	upper members	130-280	red and purple siltstone and fine ss
TRIASSIC	Chinle Fm	Moss Back M	0-170	
TRIASSIC	Chinle Fm	"Temple Mtn" M	0-100	red and green muds
TRIASSIC	Moenkopi Fm	upper member	430	reddish brown to tan siltstone and silty sandstone
TRIASSIC	Moenkopi Fm	Sinbad Ls M	50-150	*Meekoceras*
TRIASSIC	Moenkopi Fm	lower memb	140-200	green pyritic silts

Figure 7.2. Stratigraphic section of formations in the San Rafael Swell area (from Hintze, 1973).

Figure 7.3. Upper Morrison Formation capped by prominent massive Buckhorn Conglomerate which is the basal member of the Cretaceous Cedar Mountain Formation. The banded Morrison beds here are thought to have been deposited in lake environments.

San Rafael Swell is exposed with Navajo Sandstone forming the white outcrops.

5.9 Nodular gypsiferous units of the Summerville Formation exposed in the bluff (fig. 7.4) and road cuts to the west (right). Recent sand dunes blanket most of the valley bottom and alluvial flats to the south and southeast and up the flank of the hill.

Figure 7.4. Westward from the highway at approximately Mile 5.9 to well-bedded cliff-forming Summerville Formation capped by a slope zone and ledges of the lower part of the Morrison Formation.

7.3 Rise to the top of the road cuts. Summerville ledges, capped by Morrison beds, are exposed to the east. The eastern Reef of the San Rafael Swell is visible to the west and is cut with many deep, narrow, joint-controlled pockets. The dark-colored flatirons at the base of the white Navajo Sandstone are in basal fossiliferous limestone units of the Carmel Formation. The valley to the west is carved in large part in upper beds of the Carmel Formation and lower beds of the Entrada Formation.

10.2 Scenic turn out. Book Cliffs show to the north, the La Sal Mountains to the east and the Reef on the east side of the San Rafael Swell is visible to the west (fig. 7.5). Terraces of gravel-capped pediments over the Carmel-Entrada sequence are also visible in the intermediate distance to the west. These terraces are somewhat akin to pediments over the Mancos Shale in the vicinity of the Book Cliffs to the north.

10.8 Gorges through the Reef on the east flank of the San Rafael Swell are visible

Figure 7.5. View northwestward from the scenic turnout at Mile 10.2 of The Reef of the San Rafael Swell. The steep cockcombs are in Wingate and Navajo Sandstones and low V-shaped flatirons near the base are in lower beds of the Carmel Formation. The abrupt monoclinal flexure shows very well in the steep attitude of the Navajo and older beds as compared to the nearly flat lying younger series exposed toward the right.

ahead to the southwest. Much of the flat land along the road for several miles is veneered by drifting sand derived in large part from erosion of the Entrada Sandstone.

12.7 Cross Bridge Over Iron Wash. The old road shows to the east. This used to be a ford across the gully but several people have been swept downstream from here during flash floods. Road cuts are through gravel terraces related to Iron Wash. Iron Wash heads in the San Rafael Swell to the west and has cut one of the V-shaped gorges through the reef visible to the southwest. Lenticular, wavy-bedded, light green sandstone of the basal Entrada Formation to the east, is visible beneath rippled pink sandstones of the overlying part of the formation.

16.0 Drop down to south off the escarpment held up by slick rock sandstones of the Entrada Formation. The road below the escarpment is on poor exposures of the Carmel Formation which form the reddish and orange silty exposures to the southwest, beneath the light green basal Entrada Sandstone.

18.3 Poor exposures of greenish cuts in Entrada Sandstone visible along both sides of the road. The Henry Mountains are visible to the south at ten to eleven o'clock. Little Flattop, Mid Flattop, and South Flattop Buttes to the south at ten o'clock are held up by Morrison and Summerville beds.

21.1 Cross Old Woman Wash. Some light-colored Entrada Sandstone beds are exposed in the bluffs on the north and in poor exposures on the south side of the wash.

22.4 Deflation hummocks of recent sands visible both east and west of the road. These are points where sand has been trapped by growing plants and the intervening sand has been blown away, leaving hummocks of loose sand behind.

Ahead small barchan dunes are visible to the right and left of the road. These have their horns pointing toward the northeast indicating the direction of transport or origin of the sand is from the southwest.

25.2 Junction of Paved Road to the West into Temple Mountain Wash and Toward Goblin Valley State Park. There was a gas station and small store here at the junction during the uranium boom of the 1950-60s. Temple Mountain is on the skyline to the west (fig. 7.6). For a guide to Temple Mountain and Goblin Valley see Geologic Guide Segment 7a. Utah State Highway 24 continues to the south through dune fields. Several excellent ripple-marked barchan dunes are exposed in the immediate vicinity of the road.

Figure 7.6. View westward from approximately Mile 25.2 of Temple Mountain, a bleached outlier of Wingate Sandstone, which rises along the margin of the San Rafael Swell. Recent sand dunes are moving across Entrada beds.

27.7 Excellent barhan sand dunes in the immediate vicinity of the road (fig. 7.7).

28.8 Cattle Guard. Golson Butte and Little Golson Butte are ornately carved Entrada Formation and are visible on the skyline to the west. The road continues through recent sand dunes.

30.2 Panorama to the south from a high point on the road. Boulder Mountain forms the low rounded hill beyond Factory Butte to

Figure 7.7. Barchan-like sand dunes along the road at approximately Mile 27.7. Golson Butte is the high promontory beyond the dunes toward the right.

the southwest. The intricately carved green Curtis and red Entrada Formations are exposed at the south end of Goblin Valley. Thousand Lake Mountain forms the rounded promontory on the skyline at about two o'clock. High peaks of the Henry Mountains can be seen at eleven o'clock beyond Hanksville. The road is in the massive light basal Entrada Sandstone, above the red Carmel Formation.

31.8 Goblin Valley is to the west, the dark-colored Entrada beds, south of the dune patches which lap up against the ledges on this side of Wild Horse Butte. The southward gentle dip of the Entrada Formation is emphasized to the south by contact with the overlying light green Curtis beds. Summerville beds, for the most part, cap the cliffline west of the Goblin Valley area.

35.7 Emery County-Wayne County Line. Recent sand dunes are visible in the immediate vicinity particularly east of the road.

38.6 Turn out. Pointers are oriented to the major topographic features. Golson Butte is the large butte immediately west of the highway just south of the Temple Mountain area. Molly's Castle is an outlier of Entrada Sandstone which is east of Wild Horse Butte but west of Golson Butte. Monuments in the valley to the west are residual features of Entrada Sandstone (fig. 7.8). Factory Butte is visible above the gray Mancos Shale toward the southwest. Typical castellate development of the Entrada and Summerville beds show to the west. Typical Entrada stone baby beds form the butte adjacent to the highway on the east. Notice how the sand dune arches around the small erosional remnant east of the road. Wind whipping around the base of the pinnacle keeps the sand swept away.

Figure 7.8. View westward to Factory Butte from the turn out at Mile 38.6. Outliers in the foreground are in the Entrada Formation. Light-colored slope zones in the background beyond the Entrada beds are on the Curtis Formation. Slopes at the base of Factory Bette are in upper beds of the Mancos Shale and are capped by Cretaceous Sandstone.

39.6 To the east can be seen crumpled slump zone within the Entrada Formation. Most of the isolated stone babies and columns in the immediate vicinity are in intensely crumpled sections of the Entrada (fig. 7.9). Where the beds are more normally developed they erode somewhat quicker. Recent dunes blanket the Entrada Formation both east and west of the road.

42.3 Junction of the Road to the Hanksville Landing Field and Emergency Airport,

Figure 7.9. Irregularly bedded outlier of Entrada Formation at approximately Mile 39. More massive units are dolomitic and are separated by silty less resistant units which weather away to help produce the "stone babies" which are characteristic of this facies of the formation.

Figure 7.10. Westward up Muddy Creek from near the junction of Muddy Creek and the Fremont River near the bridge at Mile 43.2. Entrada beds are exposed along the inner gorge of the canyon, beneath light-colored Curtis Sandstone which, in turn, is overlain by cliff-forming, well-bedded Summerville beds. Outliers of Morrison Formation cap the ridge.

to the Left. The road descends to the south onto the floodplain of the Fremont River through a thick veneer of 20 to 30 feet of pebbly gravel terraces. The terraces are carved in the Entrada beds.

43.2 Bridge Over Muddy Creek and Fremont River. Outliers of Entrada Sandstone, to the west on both the north and south sides of Muddy Creek (fig. 7.10), are capped by green resistant beds of the Curtis Formation. High on the skyline to the southwest castles of Summerville Formation are capped by a thin basal sandstone of the Morrison Formation. The Fremont River is an overloaded stream with a rather characteristic braided channel. It is entrenched 10 to 15 feet below the general valley floor from here to west of Hanksville.

44.3 View across the river of the silty stone baby beds of the Entrada Sandstone (fig. 7.11). Notice channels are developed in the upper pinkish part of the formation. These form channel fillings from broad lenses visible from here to west of Hanksville for some distance. The road continues on the east side of the Fremont River through cuts on the Entrada Formation, all capped by

Figure 7.11. View westward of broad open minor folds in the Entrada Formation along the west side of the inner gorge of the Fremont River, from approximately Mile 45. Curtis Formation caps the valley wall.

gravel terraces. Some of the gravel terraces are up to 200 feet above the present elevation of the river.

45.7 Junction Utah State Highway 95 with Utah State Highway 24, continue on Utah State Highway 24 into Hanksville, to the west. Utah State Highway 95 leads south to Lake Powell and east to Natural Bridges National Monument and Blanding.

45.9 Entering Hanksville.

46.5 Businesses at west end of Hanksville. Reset mileage to 0.0.

0.0 Grocery store-cafe-motel and gas station in the west end of Hanksville business district.

0.3 Cross Fremont River on the New Bridge. Exposures of light-colored Curtis beds can be seen immediately to the east, west and north of the north bridge abutment.

1.3 Junction of the bypass road with the Hanksville route northwest of the Fremont River bridge. Castellate surfaces on the Summerville Formation are exposed both on the north and south sides of the river, capped by basal sandstones of the Morrison Formation.

2.0 Junction of old road with the new highway. To the south where the old bridge crosses the Fremont River early inhabitants constructed a diversion dam and reservoir but within a year the sediment-laden Fremont River completely filled the reservoir.

2.4 Excellent exposures of Summerville Formation in cuts along the cliff to the north at the south rim of North Pinto Hills. A massive conglomeratic sandstone at the base of the Morrison Formation caps the cliffs on the north and south side of the river.

3.2 Top of the Summerville Formation and base of the Morrison Formation at road level. The road continues ahead in the Morrison Formation. Manti Butte to the south is composed almost totally of Morrison Formation. Floodplain of the Fremont River is visible to the south. High gravel capped terraces associated with the Fremont River are visible north and south of the river as well as upstream to the west.

4.6 Deep double road cuts through cross-bedded sandstones of the Morrison Formation. Variegated shales of the Morrison Formation are visible across the floodplain to the south as well as near the road north of the river. Gray beds above the variegated Morrison beds are part of the Cedar Mountain Formation which are overlain by Dakota Sandstone on the skyline to the north.

6.1 Turn Out on Old Road to the North. Dakota Sandstone, filled with the oyster *Gryphaea newberryi,* occurs on the bluff to the east (fig. 7.12). This same sandstone can be seen to the north and northwest across the square-bottomed arroyo which is cut in alluvial valley fill. Oyster shells have been excavated on the terrace level here for road metal. Blocks which have tumbled down near the road are filled with the shells. The cross-bedded Carbonaceous material below

Figure 7.12. Exposures of *Gryphaea*-bearing Dakota Sandstone on the north side of the canyon at approximately Mile 6.1. Factory Butte is the chimneylike outlier in the far distance.

the Dakota Sandstone may be part of an old oxbow channel-filling.

6.5 Terrace gravels composed almost totally of oyster beds are visible on the north. These deposits are part of old oyster banks three or four feet thick in the basal unit of the Mancos Shale, possibly in a sandy beach zone at the base of the marine transgression. This is one of the famous localities for *Gryphaea newberryi.* Tununk member of the Mancos Shale is exposed in gray barren bluffs both north and south of the road with some sandy beds of the Dakota Sandstone exposed at the base.

7.0 Steamboat Butte is visible the southwest as a promontory in the Mancos Shale.

8.0 Old rock ruin of Giles or Blue Valley on the south side of the road. Mancos Shale capped by the Ferron Sandstone exposed to the north and west along the skyline rim at the east edge of Factory Bench erodes to form Lower Blue Hills.

8.6 Steamboat Point shows well to the south in front of the Henry Mountains. Smoke stacks on Steamboat Point are outliers of Ferron Sandstone.

10.1 Cross the square-bottomed arroyo. The road begins to climb to the west up through the upper part of the lower Tununk Member of the Mancos Shale into the Ferron Sandstone. That member of the Mancos Shale is exposed at road level. The road continues to climb through the Ferron Member.

10.9 Factory Butte to the north is capped by the higher Emery Sandstone. The Bluegate Member of the Mancos Shale separates the two sandstone units and form the exposures of the Upper Blue Hills visible ahead.

11.1 Coaly outcrops in the Ferron Sandstone with white bentonitic zones and lenticular cross-bedded sandstones associated with the coal.

11.8 Double road cut through the upper part of the Ferron Sandstone sequence containing a coal bed one and one-half feet thick. The coal bed is locally clinkered beneath the overlying gray middle shale of the Mancos Shale which erodes to form the Upper Blue Hills.

12.5 Badlands of the east face of North Caineville Mesa can be seen to the northwest. The flat in the foreground is a result of sheet wash. The whole surface runs when strong rains occur.

14.8 Boulder-protected pinnacles are visible in the fill above the Mancos Shale in the middle of the bluffs to the south on South Caineville Mesa.

15.4 Lathlike gypsum crystals make the gray cliff sparkle to the north. Terrace development continues on both sides of the river valley although from the highway terraces are most visible on the south side. Rocks in the general vicinity here are nearly flat lying in the trough of the Factory Butte Syncline, but dip back toward the east along the east side of the Waterpocket Fold in the vicinity of Caineville ahead.

17.4 View of the sediment-choked Fremont River visible to the south with its rather characteristic braided channel. Reversal of topography of sediment cover can be seen to the south. The ridges are now armoured by slope wash which originally accumulated in gullies. Because of resistance to erosion the debris protected the gully bottom while the bordering areas were eroded away more rapidly. What is now the ridge crest originally was the gully bottom. Individual boulders are also protecting pillars in the slope wash zone (fig. 7.13). Badlands near the road are in the Bluegate Shale. Emery Sandstone caps both North and South Caineville Mesas.

18.4 Road Junction in Caineville. Foundations for the store and church on the south side of the road. The beds now dip eastward

Figure 7.13. Hoodoos protected by boulder debris are carved in Mancos Shale on the south side of the Fremont River Valley in South Caineville Mesa at approximately Mile 14.8.

Figure 7.14. View southward along the Caineville Reef, a cuesta exposing Mancos Tununk Shale beneath a cap of resistant Ferron Sandstone. Slightly sandy beds within the Mancos Shale produce the repetitious saw-toothed resistant unit midway up the slope.

rather steeply as the road goes through the basal sequence of the Bluegate Member of the Mancos Shale very quickly. The upturned sandstone visible directly ahead are the Ferron Sandstone, this is the coal-bearing sequence seen a few miles to the east.

18.8 Top of the Ferron Sandstone in the cuesta that forms North Caineville Reef to the north and Caineville Reef to the south of the watergap of Caineville Wash.

19.5 View south along a cuesta of remarkably repetitive form in the basal Tununk Shale Member of the Mancos Shale along the west face of Caineville Reef (fig. 7.14). Dakota Sandstone is missing in this general vicinity and the basal Mancos Shale and the oyster beds rest directly on the Cedar Mountain Formation.

20.5 Summit. Ferron Sandstone forms the cuesta of Caineville Reef to the east. Mancos Shale is exposed in the immediate vicinity of the road. Gray and pinkish beds on the skyline to the west are in the Cedar Mountain and Morrison Formations (fig. 7.15).

21.5 Gray beds exposed immediately west of the road are the uppermost beds of the Cedar Mountain Formation. Variegated beds below this can occasionally be seen in gullies, interbedded with pink and brighter colored units. These colored beds are in the Morrison Formation. The Ferron Sandstone continues to form a prominent cuesta of Caineville Reef on the east side, with high country east of Caineville and the Fremont River capped by a massive sandstone of the Emery Member.

22.3 Cross the Fremont River. Exposures of Cedar Mountain and Morrison beds occur to the north and west in the Caineville Dome.

22.5 This is the Old Townsite of the Village of Caineville, first established when pioneers moved into the area. One night a flash flood almost destroyed the entire townsite along with the agricultural lands in the immediate vicinity. This destruction prompted the settlers to move the village downstream to the present site on Caineville Wash. During the flood many people were happy to escape with their lives and lost

CAPITOL REEF AREA

Age	Unit	Thickness	Remarks
TRIASSIC CONT'D	Chinle Fm — upper member	90-195	variegated silt, ss, ls
	Chinle Fm — middle member	255-345	variegated clays, *fossil wood*
	Chinle Fm — Shinarump M	0-90	ss, cg
	Moenkopi Fm	500-775	reddish brown mudstone, siltstone ripple marks are common
	Moenkopi Fm — Sinbad Ls M	70-140	*Hemiprionites*
	Moenkopi Fm — lower member	50-110	reddish siltstone
PERMIAN	Kaibab–Toroweap Ls undifferentiated	250-350	calcareous siltstone cherty ls, dol, ss
	Coconino Ss (=Cedar Mesa Ss)	800	white to gray cross-bedded ss; beds beneath Coconino are not exposed but have been penetrated by a few wells
	Elephant Canyon Formation	250-700	
PENNSYLVANIAN	Hermosa Group — Honaker Trail Formation	200-400	
	Hermosa Group — Paradox Fm	400-1000	
	Hermosa Group — Pinkerton Trail Fm	100-300	
	Molas Fm	0-150	
MISS	Redwall Ls	940	
DEV	Ouray Ls	140	
	Elbert Fm	510	
CAMBRIAN	Cambrian carbonates undivided	1200-1600	"Lynch" Dolomite; "Muav" Limestone
	Ophir Formation	200—300	
	Tintic Quartzite	200+	
P Є	Precambrian	—	

Age	Unit	Thickness	Remarks
Q	Pinedale? stage; Bull Lake? stage; Pre-Bull Lake? stage		till, outwash, gravel, boulder deposits
TERTIARY	volcanic rocks	500+	Basalts with inter-bedded tuffaceous sediments
	Flagstaff Ls	500+	*fresh water clams and snails*
	Upper Mesaverde and North Horn concealed here		
CRETACEOUS	Mesaverde Group	300+	yellow ss
	Mancos Shale — Masuk Member	600—800	sandy shale
	Mancos Shale — Emery Ss M	250	thin coal
	Mancos Shale — Blue Gate Shale Member	1400	blue-gray shale
	Mancos Shale — Ferron Ss M	250	thin coal beds
	Mancos Shale — Tununk Sh Member	575	*Gryphaea newberryi*
	Dakota Ss	0-50	
JURASSIC	Morrison Fm — Brushy Basin Ss M	60-225	variegated clays
	Morrison Fm — Salt Wash Ss	30-235	cg, ss, silt
	Summerville Fm	200	chocolate muds
	Curtis Fm	0-80	gypsiferous
	Entrada Ss	475-780	reddish brown ss
	Carmel Formation	310-990	gypsum; *Pentacrinus*; *Camptonectes*; *Trigonia*
TRIASSIC CONT'D	Navajo Sandstone	800-1100	white crossbedded ss
	Kayenta Fm	350	lenticular reddish brown sandstone
	Wingate Ss	320-370	sandstone

Figure 7.15. Stratigraphic section of rocks in the Capitol Reef Area (from Hintze, 1973).

almost all of their property into the Fremont River as the increase in water volume changed the meander pattern of the river.

Basal beds of the Mancos Shale, with *Gryphaea newberryi* beds, are exposed above the light-colored ashy Cedar Mountain beds at the south edge of the old Caineville townsite. The road pulls away from the Fremont River again and climbs up through a low pass in the lower beds of the Mancos Shale.

23.7 Junction of ranch road northward into the Fremont River. Morrison beds are exposed to the northeast. The small Caineville Dome is located north of the river and has accentuated the general eastern dip of the beds off the Waterpocket Fold. North Blue Flats north of the river are in Tununk Shale in a syncline between the Cainevile structure and the major monoclinal fold of the Waterpocket structure.

24.6 Exposures of the basal beds of the Mancos Shale in road cuts with gray ashy Cedar Mountain beds visible to the west. Occasionally thin lenses of Dakota Sandstone occur at the horizon.

25.0 Road Junction South to Notom and the general area of the Waterpocket Fold west of the Henry Mountains. **Continue ahead on Utah State Highway 24.** Road cuts on the curve are cut into ashen gray and light purple beds in the Cedar Mountain Formation. Black basalt boulder rubble veneers the general surface. These boulders were brought in from the Thousand Lake Mountain area to the west.

25.4 Junction of small road leading down to Fremont River Valley to the north. The bright-colored beds at about this point are in the Morrison Formation. The prominent sandstone conglomerate ledge approximately halfway up the slope is the Buck Horn Conglomerate, the boundary marker between the Morrison Formation and the overlying Cedar Mountain Formation. Continue ahead on Utah State Highway 24 through Morrison Formation in double road cuts.

26.0 Bridge over Pleasant Creek, a stream coming in off the flank of the Water Pocket past Notom to the south. Morrison beds are exposed on both sides of the river, with Cedar Mountain beds forming the highest parts of the exposures on the skyline around the south rim of North Blue Flats.

28.4 Top of the Summerville Formation and the base of the Morrison Formation at road level. Exposures of the crinkly gypsiferous upper Summerville beds can be seen southeast of the road and ahead for some distance. Summerville and Curtis beds erode away easily to form subsequent valleys. South Desert Valley to the northwest is typical of this development.

28.7 Junction to Notom to the South. The road to Notom climbs through the Curtis beds which are not well exposed at river level near the junction, but are well exposed just west of the junction.

29.0 Exposures of the upper Entrada Formation both north and south of the road. This is more silty and shaly than the stone baby beds where the formation was last seen near Hanksville, but it still has the same characteristic dusty red brown tone. The valley of the Fremont River widens here where the Entrada, Curtis and Summerville beds are exposed because these rocks are more easily eroded than the resistant sandstones of the overlying Morrison Formation or the underlying Carmel Formation.

29.5 Section of the road heading toward the north provides a view into the valley of South Desert along Deep Creek. The lower part of the pink section in the valley to the north is in the Entrada beds which are overlain by the prominent, but thin, light greenish band of the Curtis Sandstone. Upper pink slopes and castellate surfaces are in the Summerville Formation which is capped by massive white sandstone of the Morrison Formation.

30.4 Narrows in the canyon cut in the middle part of the calcareous Carmel Formation. Some gypsiferous units and some calcareous sandstones are more resistant in restricted beds.

31.2 Pull Out at the Waterfall. The Fremont River used to flow around the meander to the south but when the highway was constructed, the river was displaced into its present new channel through the cut (fig. 7.16). To the south the prominent pink band marks the top of the Navajo Sandstone and the base of the ledge-forming Carmel Formation above. Boulders of basalt mark the prominent gravel terraces on top of the white, cross-bedded Navajo Sandstone. To the west the massive jointed Navajo Sandstone is capped above the prominent thin bedded dark red zone by the Carmel Limestone. Some terrace gravels show at intermediate levels and at approximately the same level as the

Figure 7.16. View westward along the diverted Fremont River near a small waterfall at Mile 31.2. Massive Navajo Sandstone forms the vertical-walled inner gorge and well-bedded Carmel Formation forms the skyline rim.

terraces in the Navajo Sandstone directly to the southeast.

32.0 Elijah Cuttler Behunin Cabin. It was erected here in 1892 and was occupied by a family of nine. It is approximately an 18 × 18 foot stone structure.

32.3 Entering Capitol Reef National Monument. Massive cross-bedded Navajo Sandstone forms the outcrops on both sides. On some the irregular flat stream terraces of black basalt boulders still locally veneer the bedrock.

33.4 Grand Wash enters the Fremont River from the west. Grand Wash is one of the few gorges through which one can ride a horse or drive a jeep through the Capitol Reef. A trail leads up Grand Wash slightly over a mile to the end of the road in from the upper end. Cross-bedded Navajo Sandstone is well exposed on all sides.

34.5 The shear wall on the south is carved in Navajo Sandstone. Excellent cross bedding is visible in the sandstone on the north.

34.9 Top of the Kayenta Sandstone. The Kayenta Formation is a relatively well-bedded unit in contrast with the overlying massive Navajo Sandstone. In addition it has interbedded red shales that spill over the light-colored sandstones and stain them.

35.5 Cave forming by decementation can be seen on the left in the Kayenta Sandstone (fig. 7.17). The honeycomb development is rather common in the Kayenta Sandstone and shows well here.

36.0 Cohab trail enters on the Kayenta Sandstone ledge from the southwest. Cohab trail is maintained by the park and leads through Cohab Canyon slightly over one mile to the campground near Fruita. **Cross the Fremont River.**

36.2 Parking Area for Hickman Arch and Cohab Canyon Trail. A trail leads to the northwest to Hickman Natural Bridge approximately three-fourths of a mile. Whis-

Figure 7.17. Westward along the Fremont River Valley and the monument access road from approximately Mile 35.5. Kayenta beds form the vegetated lower slopes beneath massive rounded exposures of Navajo Sandstone. The top of the Wingate Formation is essentially at river level.

key Spring is approximately one mile on a trail and Rim Overlook is one and three quarters of a mile on the trail up a side canyon to the north and northwest. Top of the Wingate Sandstone appears south of the parking area and bridge near where the Cohab Trail comes onto the road. Massive sandstones on either side of the river in the immediate vicinity of the parking lot are upper beds of the Wingate Sandstone.

36.7 The massive shear wall of bright-colored Wingate Sandstone to the south is stained various colors from the overlying Kayenta beds. Top of the Chinle Formation and base of the angular-jointed Wingate Sandstone cliff is exposed near road level.

36.9 Petroglyph turn out. The Indian drawings, or petroglyphs, are carved in the desert varnish at the base of the cliff of the Wingate Sandstone. The large trees in the immediate vicinity were probably planted around one of the old homes which apparently was removed for construction of the road.

37.3 Capitol Reef Lodge across the gully of Sulphur Creek to the south. The road continues through fruit orchards. The lodge can be reached by turning left at the Visitors Center and return to the east on the south side of Sulphur Creek. Chinle beds are well exposed directly ahead as the lower gray green bentonitic ashy slope, a middle reddish brown slope, and an upper gray slope beneath the massive Wingate Sandstone cliff.

37.6 Base of the gray Chinle beds and top of the slabby red Moenkopi Formation on the north side of the road. Most of the rocks ahead and on the south side of the canyon of Sulphur Creek in the Moenkopi beds which have a prominent cap of basalt boulders. The road continues ahead with exposures of Moenkopi at road level. The Castle is the bold promontory ahead to the northwest (fig. 7.18).

Figure 7.18. View northward from near the Capitol Reef Monument headquarters at Mile 38.1. Moenkopi beds form the lowe prominent nearly vertical wall beneath slopes on the Chinle Formation. Strongy-jointed Wingate Sandstone forms the cliffs and the tower beyond.

38.1 Junction of the Road to the Campground and the Visitors Center. Turn Left Off Utah State Highway 24 into Visitors Center Parking Area. The Visitors Center is constructed of slabs of Moenkopi Sandstone. The old center is the more evenly-bedded rock house behind the new structure. Excellent ripple marks, rain drop impressions, mud cracks can be seen on many of the blocks in the new building. Continue west on Utah State Highway 24 (fig. 7.19). For a side trip guide in the park area to the south, see Segment 7B.

40.7 Panorama Point provides long distance views of Capitol Reef. High peaks of the Henry Mountains can be seen to the east through the gorge of Fremont Canyon.

42.7 The inner gorge of Sulphur Creek can be seen down tributary canyons to the south. The tan ledge on the rim is in the Sinbad Limestone, slope zone is in the basal red member of the Moenkopi. The shear, light-colored walls of the gorge are in the Kaibab Limestone.

43.2 A long descent on the highway with prominent white band of Sinarump Conglomerate exposed ahead between the Chinle and Moenkopi beds.

43.7 Turn out to the south offers a panorama view of the brilliant red walls of Wingate Sandstone to the north. The road continues ahead in the Moenkopi Formation. Prominent white ledge low on the north is the Shinarump Conglomerate. Trace the Shinarump ledge along and notice how lenticular it is. In some places it thickens to as much as 70 or 80 feet but in other places it virtually disappears between the red Moenkopi beds and overlying greenish Chinle Shale (fig. 7.20).

45.4 Twin Rocks. These are outliers of Shinarump Conglomerate on top of the Moenkopi Formation.

46.6 Cattle Guard at Capitol Reef Monument Boundary.

47.7 Cross Sulphur Creek. Swampy areas here are probably the result of irrigation of the terrace country to the west.

50.3 Junction of Utah State Highway 117 Southward to Escalante Over Boulder Mountain. Continue Straight Ahead on Utah State 24 Toward Torrey and Bicknell on the north side of the Fremont River valley. Navajo and Carmel beds can be seen south of the Teasdale Fault, beneath the volcanic cap and boulder debris of Boulder Mountain across the valley to the south.

50.9 Enter Torrey. Thousand Lake Mountain is at two-thirty to the north. The LDS Churchhouse at Torrey is constructed out of red sandstones from the Moenkopi Formation. Coarse boulders in the vicinity of Torrey are debris of flash floods and mudflows along Sand Creek from Thousand Lake Mountain on the north. These debris fans blanket a pediment cut in Moenkopi beds.

52.1 Coarse lava boulder rubble on Poverty Flat, on the slope of the alluvial fan, out of Sand Creek covers Moenkopi rocks.

53.2 Cross the Fremont River. The river is entrenched 20 to 30 feet here below the general gravel-capped terrace surface. The road rises to the terrace surface which is formed both north and south of the road. Moenkopi and Shinarump beds form the fluted wall rim of Velvet Ridge to the north, just beyond the valley.

54.8 Junction of Utah State Highway 54 from Teasdale to the South with Utah State Highway 24. To the southwest a nearly complete but somewhat faulted sequence of Moenkopi Formation to Carmel Limestone is exposed in the partially debris covered hills. Hummocky areas south of Teasdale are in landslide debris.

56.6 Cross the Fremont River. The old mill on the north used to have a water wheel on the west side. The millrace is the ditch crossed by the road just after crossing the Fremont River. Shinarump Conglomerate is exposed on the ridge north of the road.

56.9 Entering broad Rabbit Valley at the

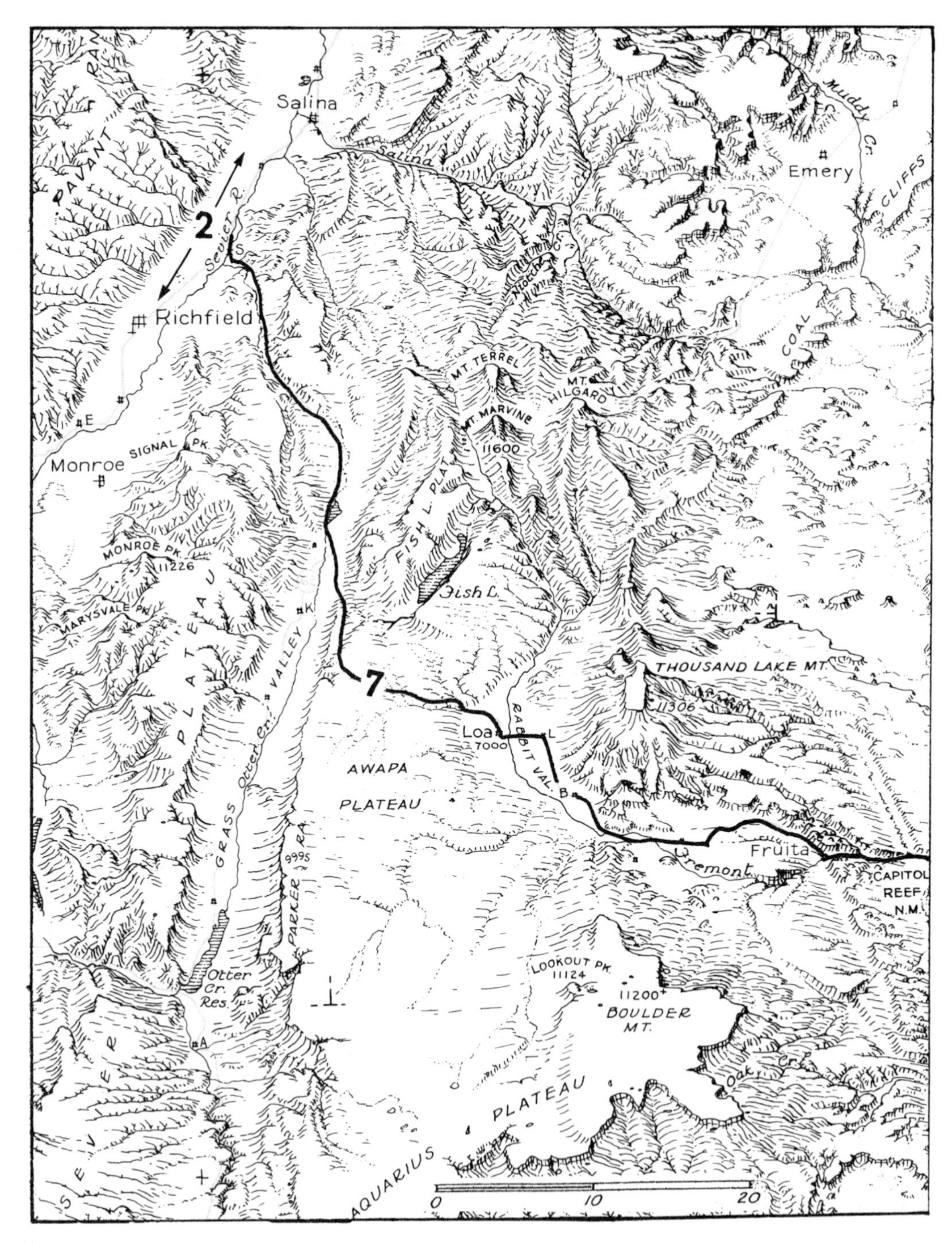

Figure 7.19. Index map to the western part of Route 7 from near Capitol Reef National Monument to the end of Route 7 where it joins Route 2 near Sigurd.

Figure 7.20. Banded castellate cliffs of upper Moenkopi beds beneath Shinarump and Chinle rocks along the north side of the highway at approximately Mile 44.

upper end of the Fremont drainage and cross the trace of the Thousand Lake Fault. Rabbit Valley to the west is on the downdropped block. Bicknell Bottoms are located to the west. The broad high lava-covered slopes of Fish Lake Plateau can be seen across the valley to the west. Lava-capped terraces still spill over Moenkopi and Carmel beds on the east side of the road in the immediate vicinity.

58.1 Exposures of the red rocks of the Moenkopi up to the pink Carmel Limestone can be seen east of the Thousand Lake Fault on the western margin of Thousand Lake Mountain. Alluvial fans from the mountain are deeply blanketed with black lava boulders. The boulders occasionally accumulate a light gray calcareous cover after they have been partially buried in soil. Where they have been turned over to expose this cover on their lower surface, such as along the highway, they appear black and white.

59.6 Entering Bicknell. Gray debris and volcanic rocks of the Thousand Lake Mountain complex are visible to the northeast of town, beyond ashy outcrops of Tertiary Flagstaff or Brian Head-equivalent rocks in the immediate bluffs. Basalt caps the summit.

64.2 Enter Lyman. Volcanic exposures and debris from the Thousand Lake Mountain area blanket the mountain flank east of Lyman. Immediate exposures are ashy Paleocene Flagstaff Limestone or Brian Head equivalents.

66.7 Junction Utah State Highway State 250 with Utah State Highway 24. State Highway 250 leads north from the middle of the valley to Fremont at the north end of the valley. Continue ahead on Utah State Highway 24.

67.9 Enter Loa.

69.2 Junction of Utah State Highway 72 Northeast to Fremont. Continue on north and northwest of the junction on Utah State Highway 24 toward Fish Lake, Sigurd, and Richfield. Upper end of the valley to the north is fault controlled, with lavas exposed on both sides of the valley. The road begins to climb to the northwest over the low foothills of the Awapa or Fish Lake Plateau.

71.4 Center of a long swinging bend in the road with bouldery debris from the Fish Lake volcanic field exposed in road cuts.

72.6 Outcrops of Early Tertiary basalt, andesite, and basaltic andesite can be seen in the deep V-shaped gorge cut in the pediment surface to the south.

73.4 View at seven o'clock to the southeast shows typical irregular ragged ledge of Early Tertiary volcanic outcrops in the V-shaped canyon which has entrenched into the general upland surface of the Awapa Plateau. The road continues ahead on a terrace with the entrenched V-shaped canyon to the south. Tertiary volcanic material is exposed on the bluff to the north. In general, however, the volcanic rocks here form poor exposures, presumably because of the very deep weathering on the uplands of the plateau.

77.3 The wide notch on the skyline to the north is caused by the downdropped graben block of the Fish Lake Area.

80.3 Summit-Piute County-Wayne County Line. Elevation 8,406. Fish Lake Mountain forms the prominent skyline knob to the north. The sharp indentation of the skyline to the northeast is the Fish Lake graben. Directly to the west is Koosharem Valley directly to the west in front of Koosharem Mountain on the skyline. The area to the west is also a part of the Fish Lake volcanic providence but has been faulted. The road descends the western fault line and monclinal scarp into Koosharem Valley. Exposures of deeply weathered volcanic debris and tuffaceous rocks occur in road cuts.

81.5 View into Koosharem Valley and over the community of Koosharem to the west. The escarpment here is probably a long fault zone which is coupled with some monoclinal downbending, structure is uncertain, the bedding in the volcanic sequence is poor, and key beds are rare.

82.3 Junction of Utah State Highway 25 to Fish Lake. Continue Straight Ahead on Utah State Highway 24. The road into Fish Lake climbs the escarpment.

84.4 Sevier County-Piute County Line East of Summit. A cuesta of volcanic rocks rises to the west of the road. The road is on a less resistant unit of the Oligocene Bullion Canyon volcanic sequence east of the Paunsaugunt Fault.

89.1 Junction of Utah State Highway 62 to Koosharem with Utah State Highway 24. Continue Straight Ahead on Utah State Highway 24.

90.8 West abutment of the dam for Koosharem Reservoir. High peaks of the Fish Lake Plateau visible to the east across Koosharem Reservoir have been glaciated, although the evidence for glaciation can't be seen from here. The road continues to the north along the west side of the valley. The gentle eastward slope is on Bullion Canyon volcanic debris west of the Valley Fault. The entire country here has a rather somber gray tone which is because of the purplish gray and dark brown volcanic rocks that produce most of the topography.

97.2 Junction of Forest Service Road to the north into the upper end of Koosharem Valley. Continue on Utah State Highway 24. The road swings to the west through a low pass north of Cove Mountain through the Bullion Canyon volcanics.

99.3 Beginning of Steep Downgrade into Peterson Creek Canyon off the north flank of Cove Mountain. Junipers and sage brush have been railed off here to improve the range character and productivity. Bedrock still consists of deeply weathered volcanic debris.

100.2 Peterson Creek is an entrenched small stream which is 10 to 20 feet below the projected grade of the terraces on both sides. Such entrenching is common over much of the Fish Lake Plateau and the other high plateaus of central Utah.

100.7 Cross Peterson Creek. Weathered volcanic rocks are exposed in road cuts just beyond the bridge abutment and in ragged exposures on both sides of the canyon wall above the flat valley fill on down the canyon.

101.9 Weathered lava flows and debris exposed in road cuts. These are probably the best exposed volcanic rocks along the highway.

105.9 Junction of Utah State Highway 119 West to Richfield with Utah State Highway 24. Continue Ahead on Utah State Highway 24. Road side exposures are in weathered volcanic debris and tuffaceous material. The road continues to the north in debris from the Tertiary volcanic field to the east.

106.5 Cross the bridge. Exposures of Jurassic Arapien Shale are visible directly ahead. This formation is equivalent to the Carmel Formation exposed in the San Rafael Swell, Capitol Reef Area, and country to the east of the Wasatch and Fish Lake Plateaus. To the east Tertiary volcanic rocks erode to form cliffs of hoodoos and less severe slopes (fig. 7.21).

Figure 7.21. Eastward across a small valley carved in Arapien Shale at approximately Mile 106 to irregular hoodoos and erosional remnants in volcanic rocks along the western margin of the Fish Lake Plateau.

107.4 Arapien Shale is exposed in contorted and salt-heaved outcrops with massive gypsum beds in light gray, gray green, and pink shales exposed on both sides of the valley. Outcrops to the southwest are partially buried beneath terrace gravels of volcanic debris.

108.7 Massive gypsum of the Arapien Shale is exposed in road cuts on the east side of the road. Koosharem Mountain and the Pavant Range behind Richfield can be seen over the low hills of Arapien Shale to the west at ten o'clock. The road continues through exposures of Arapien Shale (fig. 7.22) with the valley floor filled with alluvium. Regional dips are difficult to obtain in the Arapien Shale because of complex heaving and tectonic involvement.

Figure 7.22. Crumpled and folded gypsiferous Arapien Shale exposed on the north side of the canyon at approximately Mile 108.7. More resistant units are gypsum beds interspersed in the easily eroded silty shale.

110.1 Cross a small bridge and continue on down into the narrows of the lower part of the canyon. Conglomeratic terrace gravels are exposed on the north side of the road. Some of these deposits have been used for road metal.

111.3 Cross trace west of the Sevier Fault at the mouth of the canyon. Plasterboard plants are visible to the north. These plants utilize gypsum from the Arapien Shale exposed along the valley margin for some miles to the northeast.

112.1 Intersection in Sigurd. The U.S. Gypsum Plasterboard Plant is directly to the east. Continue across the railroad tracks. North out of Sigurd on Utah State Highway 24.

112.7 Meander of the Sevier River. The river is backed up into Rockford Reservoir by a low dam downstream a short distance. Red rocks to the west, across Sevier Valley are part of the Paleocene Flagstaff or the Bryce Canyon Formation. This is about the southernmost exposure of the formation locally for south of Richfield it is covered by volcanics of the Marysvale district and Fish Lake Plateau.

113.4 Entering Vermilion. Vermilion received its name from the bright red soil and debris brought into the Sevier Valley by streams draining the Pavant Range to the west. Excellent exposures of massive gypsum in the Arapien Formation can be seen in the west face of the Fish Lake Plateau to the east. A complex road and mine system of the gypsum mines is mainly along the low flank of the plateau in the valley behind the small foothills.

114.0 Cross Over Tracks of D&RGW Railroad and Junction with U.S. Highway 89. For a description of the route to the north and south see Geologic Guide Segment 2.

Segment 7A

0.0 Turn to the Right Toward Temple Mountain. Sand dunes are visible to the north and south of the Temple Mountain Wash Road. The road continues over poor exposures of Entrada Sandstone, now largely buried by thick drifting belts of sand.

1.2 Gilson Butte and Little Gilson Butte to the south are in the stone baby beds of the Entrada Formation. Temple Mountain is visible on the skyline to the northwest at about two o'clock. It is a bleached outlier of Wingate and Navajo sandstones. Much uranium was produced in the immediate vicinity of Temple Mountain.

4.5 Exposures of white sandstone and gray units of the Entrada Sandstone form eastward dipping cuesta.

5.3 Junction of Goblin Valley Road to the South with the Temple Mountain Road. Continue Ahead on Paved Road into the Temple Mountain Uranium District.

5.9 Entering the San Rafael Reef of massive white cross-bedded sandstones of the Navajo Formation along South Temple Wash. Differential erosion has produced a somewhat pitted surface on the Navajo Sandstone. The prominent joints along which streams have cut outward from the San Rafael Swell can be seen ahead.

6.4 Top of the Kayenta Formation. This is the white flaggy-bedded sandstone. In the whole zone here the rocks have been bleached a light gray.

6.7 End of the paved road. Top of the Wingate Sandstone. The Wingate Sandstone has also been bleached to a light color. This is an unusual color for the formation for it is usually a brick red angular weathering formation.

6.9 Base of the Wingate Sandstones and top of the red Chinle beds is exposed both north and south of the road. Excellent exposures of sandy Chinle beds to the south, with variegated shales particularly well exposed in the lower part of the formation.

7.3 Cross through the upper beds of the Shinarump Conglomerate. Temple Mountain Wash uranium mines and glory holes can be seen toward the north toward Temple Mountain, the prominent white peak on the skyline. Glory holes in the near exposures are in the Shinarump Conglomerate.

7.7 Excellent exposures of Moenkopi beds both north and south of the road at a gully crossing. This road continues on through the San Rafael Swell and ultimately connects with Interstate 70 in the center of San Rafael Swell and with the road from Castle Dale down Buckhorn Wash across the northwest part of the dome.

Turn around in the forks in the road and return to the junction of the Goblin Valley Road at Mile 5. 3 An excellent view of Temple Mountain to the north (fig. 7.23) can be seen from just west of the old Temple Wash uranium campsite. As late as 1968 there were

Figure 7.23. View northward to Temple Mountain from near Mile 7.7. Rocks in the foreground are Moenkopi Formation. Chinle beds form the light-colored slope and ledge zone in the background and are capped by outliers of bleached Wingate Sandstone which forms Temple Mountain.

still some leasor operations in mines on the Shinarup Conglomerate about halfway between the campsite and Temple Mountain. Navajo Sandstone forms the jointed upper surface of the cliff to the east with bleached Kayenta and Wingate Sandstones exposed on the south side of the gorge.

9.8 Junction of the Goblin Valley Road at Incoming Mile 5.3. Turn to the South (Right). The road continues to the south through gravel-capped terraces and sand dunes, toward Goblin Valley. The white Navajo Sandstone Reef is visible to the west.

10.7 To the southwest excellent exposures of Carmel Formation can be seen lapping up onto white sandstone outcrops of the Navajo Sandstone. Kayenta beds form the juniper-covered skyline exposures. In the far distance, at two o'clock, Boulder Mountain forms the skyline. Factory Butte forms the square-topped badland butte to the southwest in the distance. Wild Horse Butte and the edge of Goblin Valley are carved in the Summerville, Curtis, and Entrada beds in the near distance at one o'clock.

11.8 The road drops off the gravel terrace into exposures of the Carmel Formation.

12.8 Road Junction to Buckskin Springs. Reddish Carmel beds are well exposed along both sides of the road beneath the light gray basal sandstone of the Entrada Formation.

14.0 Crinkling in the Carmel Formation can be seen in the canyon bottom to the west. This may be related in part to salt heaving and in part to differential sedimentary loading during deposition. The south edge of the San Rafael Swell can be seen to the southwest. Thousand Lake Mountain forms the skyline in the distance.

15.1 Junction of Road to the Southwest Toward Wild Horse Spring with the Main Goblin Valley Road. Continue ahead on main road to Goblin Valley. Wild Horse Spring is in Wild Horse Wash to the west.

Molly's Castle viewpoint is just beyond the junction. Molly's Castle to the southeast is a remnant of the Entrada Formation capped with the greenish Curtis beds. Basal reddish castellate surfaces are in the Entrada beds. The road descends through light greenish tan cross-bedded Entrada Sandstone which forms a slick rock exposure. The beds are dipping steeply to the southeast here and are cut by numerous small faults.

16.1 Junction of Goblin Valley Road with the principal road to Wild Horse Mesa and the Muddy River around the south end of the San Rafael Swell. **Keep Left Around the Southeast Side of Wild Horse Butte (fig. 7.24).** Wild Horse Butte is in the red Summerville, light green Curtis and upper reddish castellate stone baby beds of the Entrada Formation. Crinkly laminated evaporitic Carmel beds can be seen in gullies to the northwest.

17.5 Goblins and stone babies are in the Entrada and can be seen ahead and to the

Figure 7.24. View southeastward to Wild Horse Butte which exposed Entrada Formation, as the lower cliff zone, and Summerville Formation, as the upper cliff zone, and Summerville Formation, as the upper cliff zone, separated by a light-colored slope carved on the Curtis Sandstone.

Figure 7.25. View southward into Goblin Valley State Park at Mile 18. The "goblins" are produced by differential weathering within the "stone baby" beds of the Entrada Formation.

left. The road leads down through massive white basal Entrada Sandstone. This is the same white sandstone which forms the slick rock exposures to the north.

17.8 The road leaves the slick rock exposures and climbs up onto the ridge at the north wall of Goblin Valley. Excellent little Goblins called the Chessmen are visible to the north and are isolated from the main valley which is south of the rim.

18.0 Road Junction. Turn South (Right) to the West, Goblin Valley Overlook. Goblin Valley can be seen both east and west from the overlook point. The road crests out at the top of the north wall of Goblin Valley at the first overlook (fig. 7.25). Goblins are formed where the beds have been jointed in resistant siltstones which alternate with thin relatively easily eroded darker shales. A jeep road leads down from the overlook into Goblin Valley and out to the south end but shouldn't be attempted without local guides. Several unimproved trails lead down into Goblin Valley. Take time to walk around on the valley floor in among the goblins (fig. 7.26).

18.1 East overlook into the north end of Goblin Valley. **Turn around and return to the Temple Mountain Road and then to Utah State Highway 24.**

Figure 7.26. Goblins of Goblin Valley. More resistant units are slightly dolomitic and withstand weathering more than do the shaly silty beds which recede to form slopes.

Segment 7B

0.0 Leave Parking Area on Paved Road to the Southeast which leads back to Fruita and the Capitol Reef Lodge and campground area on the south side of Sulphur Creek.

1.0 Capitol Reef Lodge with restaurant, rooms and a curio shop. End of the paved road.

1.1 Cross the Fremont River.

1.3 West end of the Cohab Canyon Trail. **Junction to the Campground.** The campground has a restroom and trailer facilities with lawns along the bank of the Fremont River. Excellent exposures of Chinle Shale can be seen east of the river and east of the campground. Boulder-covered terraces are developed along the Fremont River to the west. Old basalt rock fences can be seen on the river bluff. **Continue on South on the Main Park Road Through Exposures of Basal Chinle Beds.** Moenkopi rocks are exposed to the west. The Sinbad Limestone Member of the Moenkopi Formation is visible as the tan bluffs up the Fremont River to the west, above the lighter cliffs of the Kaibab Limestone. The road continues south in Moenkopi beds.

2.7 Danish Hill, summit of the divide between Grand Wash and Fremont River. The Fremont River can be seen ahead, with the prominent spur between Sulphur Creek and the river capped by dark basaltic debris.

3.0 The steep east-dipping Sinbad Limestone Member of the Moenkopi Formation forms the tan resistant ledges west of the soft reddish unit in the subsequent valley where the road has been constructed. Kaibab Limestone is exposed to the west in canyons cut through the lower Moenkopi beds.

3.6 Junction of Side Road into Grand Wash, Turn East (Left) Down Grand Wash. Cross over the approximate trace of the Shinarump Conglomerate. Prospect pits for uranium can be seen to the north in the conglomerate between the castellate surfaces of the Monekopi Formation and the slope-forming Chinle beds above. To the north the Shinarump Conglomerate pinches out a few tenths of a mile north of Grand Wash so that there Chinle beds rest directly on red Moenkopi rocks.

3.8 Cross Grand Wash. Chinle beds are exposed along both sides of the wash. Downstream well-developed terraces occur about twenty feet above the present stream channel.

4.3 Drop into Grand Wash. Uppermost beds of the Chinle Formation are preserved in vertical-walled exposures to the south and north. Top of the Chinle Shale and base of the Wingate Sandstone exposed on the north wall of the canyon ahead. Vertical walls are in Wingate Sandstone. The road continues along the bottom of the wash.

4.6 Shinob or Bear Canyon, a prominent joint-controlled canyon, joins Grand Wash from the southwest. Directly ahead the massive desert-varnished brown Wingate of Echo Cliff is capped by flaggy beds of the Kayenta Sandstone. Rounded Navajo Sandstone is visible high on the skyline.

4.9 End of Road. Parking Area. Cassidy Arch is along a foot trail approximately 2 miles from the parking area. The trail leaves Grand Wash near the Narrows and connects with the Cohab Canyon Trail approximately 4 miles from here and 1.5 miles from the campground at Fruita. A trail leads through the Narrows of Grand Wash 1.3 miles beyond the parking area. The Narrows of Grand Wash are in the Navajo Sandstone which is exposed here high on the wall to the southeast. **Turn around and return to the main road at the head of Grand Wash.** The relatively flat top of the spectacularly cross-bedded Wingate Sandstone can be seen to the west below where it grades into the flaggy beds of the Kayenta Sandstone.

6.2 Rejoin the Main Monument Road. Turn to the South (Left). The road continues southward in the soft middle beds of the Moenkopi Formation (fig. 7.27). The tan resistant unit to the west on Miners Mountain is the Sinbad Limestone Member of the Moenkopi Formation. Kaibab Limestone is exposed in the deeper canyons on the slope, below the lower Moenkopi beds, with local inliers of Coconino Sandstone in some of the deeper canyons associated with Grand Wash.

6.7 The Shinarump Conglomerate ledge to the east is about 30 feet thick, separating the gray green shales of the Chinle Formation from the brick red laminated beds of the Moenkopi Formation below. The Shinarump Conglomerate is lenticular and uneven along the outcrop trace. It disappears in the distance from the promontory back to the main body of the cliff.

7.4 To the east the massive high wall is composed of Wingate, Kayenta, and Navajo

Figure 7.27. Thin veinlets of white crystalline gypsum cutting through Moenkopi beds along the park road at approximately Mile 6.2 in Arches National Monument.

Sandstone above the slopes on Chinle and Moenkopi Formations. Lenticular Shinarump Conglomerate is now present east of the road between the Moenkopi and Chinle beds. The Shinarump Conglomerate represents old stream deposits, channel fillings or point bar deposits so by studying the unit we can get some indication of the direction of flow in the old river in which the sediments accumulated.

8.1 Bridge over gully. Excellent cross-bedded Moenkopi Sandstone is visible in the channel bottom and in abutments of the bridges. Almost all of the slabs of sandstone in this section of the formation show cross-bedding and intricate ripple marks. Jointing in the formation produces rather squarish blocks, almost ideal for building stones, that are ripple marked, ornamental and yet resistant much like those used in the construction of the Visitors Center.

8.7 Summit and Drainage Divide Between Grand Wash and Capitol Wash. Excellent high exposures of Moenkopi beds are visible to the east. The road continues to wind southward through soft siltstones and fine grained sandstones of the Moenkopi

Formation with more well-cemented beds forming semiresistant ledges. To the west occasional glimpses can be seen of the Sinbad Limestone and lighter colored sandstones in the lower part of the Moenkopi Formation exposed in gully headwaters.

10.0 Egyptian Temple to the Northeast (fig. 7.28) is carved in the upper beds of the Moenkopi Formation which are capped by the tan Shinarump Conglomerate.

Figure 7.28. View northeastward to the Egyptian Temple carved in banded Moenkopi Formation and capped by massive sandstone of the Shinarump Member of the Chinle Formation, at approximately Mile 10 in the road log.

10.6 Narrows Through Shinarump Conglomerate. This formation is a cross-bedded, stream-deposited conglomerate here and appears rather characteristic of sands that produce uranium. Grayish rounded bluffs of Chinle Formations can be seen directly ahead through the Narrows. Fossil wood has been found locally in the Chinle Formation.

10.8 Junction of Capitol Gorge road with one to the Sleeping Rainbow Guest Ranch and Pleasant Creek to the south. Continue into the Ranger Station parking area. A complete Kaibab Limestone-Lower Moenkopi Formation section can be seen by walking approximatey two miles up either fork of Capitol Wash to the west to where Coconino Sandstone is exposed in both forks of the wash.

10.9 Ranger Station and Road into the Mouth of Capitol Gorge. Excellent exhibits at the station explain the local geology and natural history. Eph Hanks Tower in Wingate Sandstone is to the south. Continue to the east into Capitol Wash, if the weather is good. **Do not go in if rain threatens, or if gates are closed because of flash flood danger.**

11.1 Tapestry walls are in Wingate Sandstone on the south wall of the gorge. Tapestrylike appearance is because of streaks of desert varnish over the Wingate Cliff.

11.5 Excellent exposures of cross-bedded Wingate Sandstone. Flaggy Kayenta rocks form the semi-slope zone in the cliff and are overlain by rounded Navajo Sandstone which caps the skyline to the east.

11.8 Wingate Sandstone forms the relatively narrow section of the canyon. Sign points to Waterpockets, or little reservoir depressions in some of the joint-controlled gullies to the south. The road continues down the bottom of the gorge cut in cross-bedded sandstone. Before the new paved road was constructed down the Fremont River Valley this was the only drivable east-west road through the Waterpocket Fold. Local ranchers tell of 10-foot walls of water in the past rushing down through the gorge during sudden summer thunder storms.

12.5 Golden Throne Parking Area. The trail leads approximately one hundred yards to the south to a view point of the Golden Throne on the skyline to the north. Top of the Wingate Sandstone and base of the Kayenta Sandstone is at road level here. Most of the rocks in the immediate canyon wall are in the Kayenta Sandstone. To see the Golden Throne walk a hundred yards down the gravelled trail from the parking lot and then look back up the canyon to the north.

13.1 Paved Crossing and Entry into the Parking Lot at the End of the Road. A walk downstream of approximately 1 mile will pass the Narrows, petroglyphs, Pioneer Register, and "water pockets" in joints in the Navajo Sandstone. The base of the Navajo Sandstone and top of the flaggy Kayenta beds come down to river level just beyond the shelter and information center at the parking lot. The contact is at the level where the flaggy reddish bedded materials give away to the light tan and white massive Navajo Sandstone. The top of the Kayenta Sandstone is exposed at river level about one hundred yards beyond the Visitors Center, just short of where typical Fremont petroglyphs are carved on the north wall of the canyon. The Narrows of Capitol Wash is in the Navajo Sandstone (fig. 7.29) approximately 1 mile below the parking area where the gorge narrows down to 15 to 20 feet wide. Navajo Sandstone in the Narrows is cross-bedded and in some beds it contains little rows of pebbly-appearing nodules. These are merely well-cemented areas in the cross-bedded fine-grained sandstone.

Figure 7.29. The Narrows of Capitol Gorge are nearly vertical walled and have been cut through Navajo Sandstone. This was the main route through Capitol Reef until a few years ago when the new road was constructed along the Fremont River. The road through Capitol Gorge was tenuous, particularly during the summer seasons of thunderstorms and flash floods.

Trails lead to the north from the parking area to where one can get a better view of the gorge and of the Golden Throne. There are interpretative exhibits, garbage facilities, and restroom facilities at the parking area. **Turn around and return to the headquarters area.**

Segment 8

0.0 Junction of Utah State Highway 15 at Interchange on Interstate Highway 15. Turn off Interstate 15 and continue eastward toward Hurricane and Zion Canyon National Park (fig. 8.1).

0.8 Double road cuts through Moenave Formation which shows some complex tight folds and minor faulting. These incompetent beds must have folded during development of the Harrisburg Anticline and other folds of the area. They have similar trends.

1.1 Junction of old U.S. Highway 89 at abandoned Harrisburg Junction. Shinarump Sandstone (fig. 8.2.) forms the checkerboard jointed surface to the south, capping a cuesta off the northwestern flank of Harrisburg Anticline.

1.4 Road cuts through Shinarump Conglomerate on the northwest flank of the fold. Cross-bedded, and swirled-bedded conglomeratic sandstone is typical of the formation. Fossil plant debris has been found locally in the unit north and south of the highway.

From an overlook south of the cuts (fig. 8.3) the northward plunging structure of the fold is clearly demonstrated by the closure of the varicolored Shnabkaib Member of the Moenkopi Formation around the northern nose. Kaibab Limestone is exposed as the resistant core of the structure to the southeast. The cuesta of Shinarump Sandstone forms the high rim around the structure, above the lower Moenkopi interior and the outer margin of Chinle beds.

Toward the east the road drops through cuts in the Shinarump beds onto brick red Upper Red Member, then onto candy-striped Shnabkaib beds in less than one-half mile.

1.8 Top of Shnabkaib gypsiferous cyclic beds. These rocks probably represent a tidal flat environment associated with a seaway which was to the west in Nevada.

3.5 Cross through Shinarump Sandstone cuesta on the east side of Harrisburg Dome and then **Cross the Virgin River** which is in a valley adjusted to the softer lower Chinle Shale on top of the resistant Shinarump Conglomerate. A short distance beyond the Virgin River bridge the highway rises onto a series of relatively recent basalt flows which occupied a former valley of the Virgin River. The river has now reexcavated a channel on the western margin of the basalt.

4.7 The road rises from the relatively older flows of the Virgin River Valley onto younger basalt flows which issued from near the small cinder cones visible ahead to the east. Road cuts along the highway are in Moenave or Navajo Sandstone beneath the basalt cap.

6.7 Volcano Mountain to the south is a cinder cone and the vent was the source for the basalt flows over which we've been riding. This and the two small cones to the

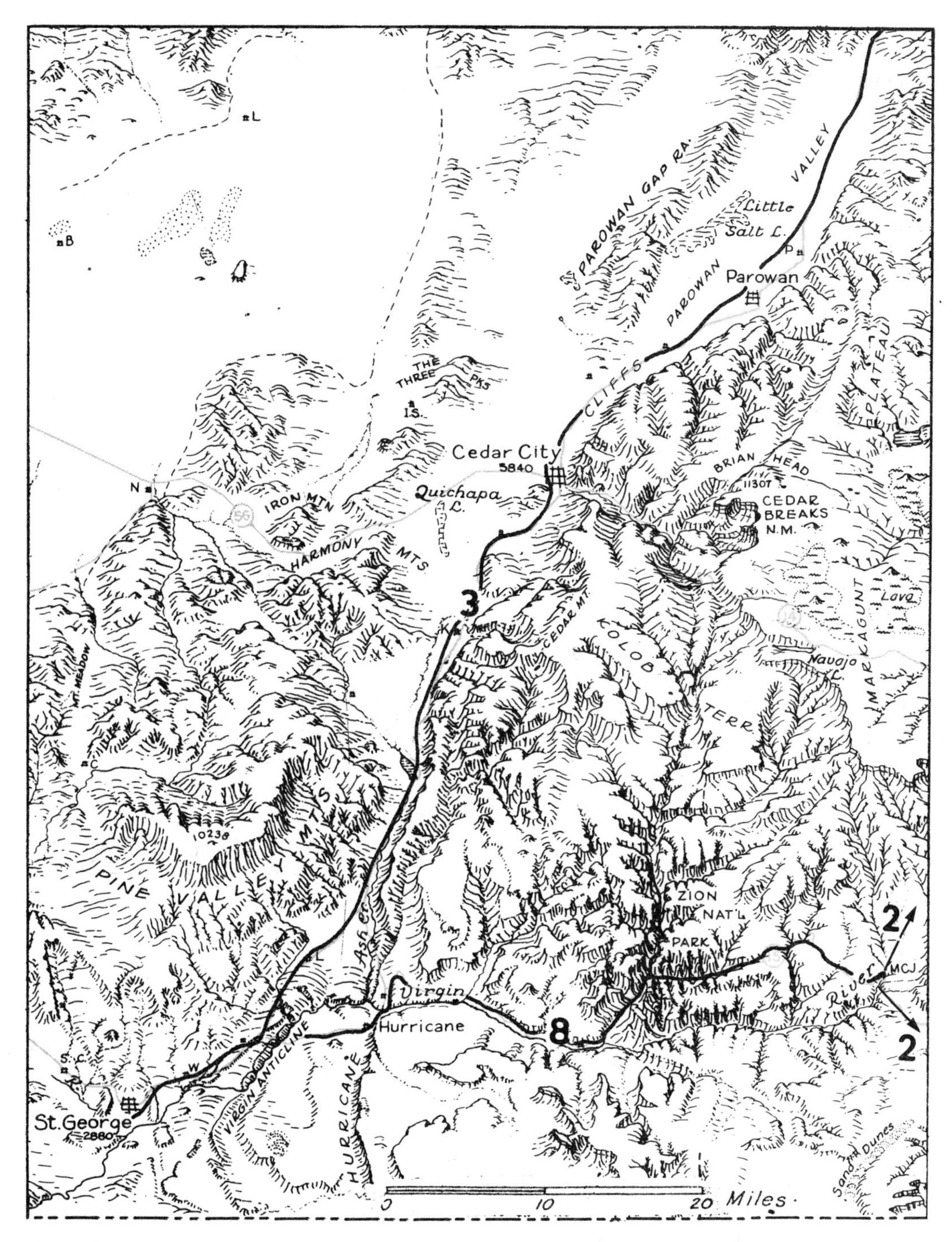

Figure 8.1. Index map of Route 8, from near St. George eastward through Zion National Park to Route 2 at Mt. Carmel Junction (MCJ) (base from Merrill K. Ridd).

SOUTHWESTERN UTAH
CEDAR CITY - ZION PARK - ST. GEORGE - IRON MINES

System	Formation / Member	Thickness	Remarks
TRIASSIC	Kayenta Fm	700–1200	red mudstone; Shurtz Ss tongue
TRIASSIC	Moenave Fm: Springdale Ss	110	red mudstone
TRIASSIC	Moenave Fm: Whitmore Pt M	15-70	pale red ss ledge; gray slope
TRIASSIC	Moenave Fm: Dinosaur Canyon M	400	light brown
TRIASSIC	Chinle Fm: Pet Forest M	250-360	*petrified wood*
TRIASSIC	Chinle Fm: Shinarump M	40-150	
TRIASSIC	Moenkopi Formation (1810): upper red	510	
TRIASSIC	Moenkopi Formation: Shnabkaib Member	320	gypsum
TRIASSIC	Moenkopi Formation: middle red	300	
TRIASSIC	Moenkopi Formation: Virgin Ls M	130	*Tirolites*
TRIASSIC	Moenkopi Formation: lower red	450	
TRIASSIC	Moenkopi Formation: Timpoweap M	100	*Meekoceras*; UNCONFORMITY
PERMIAN	Kaibab Ls	850	*Dictyoclostus*
PERMIAN	Toroweap Fm	230	*Pugnoides*
PERMIAN	Conconino-Queantoweap Ss undivided	1250	
PERMIAN	Pakoon Fm	300	*Schwagerina*
PENN	Callville Ls	900	*Triticites Fusilina Fusilinella Millerella*
MISS	Redwall Ls	960	*corals brachiopods*; cherty
DEV	Crystal Pass ? Guilmette ? Simonson ? Dolomite	800	
	Ord-Sil? dolomite	200	cherty
CAMBRIAN	Muav Ls	1100	*unfossiliferous*
CAMBRIAN	Pioche Shale	215	*trilobite fragments*
CAMBRIAN	Prospect Mtn Quartzite	530	
P€	Vishnu Schist	—	1650 M.Y. K-Ar

System	Formation / Member	Thickness	Remarks
Q	Basalt flows	0-500	6.0 M.Y. K-Ar Fortification basalt in Nevada caps Muddy Creek Fm
PLIO	Muddy Creek Fm	0-1400	
M	Page Ranch Volc	0-400	Three Peaks intrusive 21.1 M.Y.
OLIGOCENE	Rencher Fm	0-1000	21.5 M.Y.; tuff breccia, welded tuff, associated sediments
OLIGOCENE	Quichapa Fm	0-1600	22.0 M.Y.; ignimbrites; 24.0 M.Y.; 24.3 M.Y.
OLIGOCENE	Isom Ignimbrite	0-150	25.7 M.Y.
OLIGOCENE	Needles Range Tuffs	0-200	29 M.Y. K-Ar
PALEOCENE (Lower Eoc)	Cedar Breaks Fm; * max. thickness at Cedar Breaks = Claron Fm of west Utah	400– *1400	"White Claron"; "Red Claron"
CRETACEOUS	Iron Springs Formation: Kaiparowitz equiv.	4000±	
CRETACEOUS	Iron Springs Formation: Wahweap - Straight Cliffs equiv.		
CRETACEOUS	Iron Springs Formation: Tropic Shale equiv.		coal
CRETACEOUS	Iron Springs Formation: Dakota Ss equiv.		
JURASSIC	Carmel Fm: Winsor Mbr	180-320	
JURASSIC	Carmel Fm: Paria Riv Gyp M	70-150	"Curtis gypsum"
JURASSIC	Carmel Fm: Crystal Creek M	170	"Entrada"
JURASSIC	Carmel Fm: Kolob Ls Mbr	220-480	*Pentacrinus*; *clams oysters*
JURASSIC	Temple Cap Mbr		
JURASSIC	Navajo Ss	2000–2300	sand dune cross-bedding

Figure 8.2. Stratigraphic section of rocks exposed in southwestern Utah in the Zion National Park and St. George Area (from Hintze, 1973).

Figure 8.3. Northward from overlook at Mile 1.4 along the west side of Harrisburg Anticline. Shinarump Sandstone caps the slope zone on the western rim and is exposed in the foreground above various members of the Moenkopi Formation which have eroded away in the center of the breached anticline to form the lowlands.

Figure 8.4. The gorge of the Virgin River, here rimmed by partially excavated ledges of basalt, as seen westward from Mile 11.3. Basalt is on the down-dropped block of the Hurricane Fault.

north are some of the better preserved volcancoes of the state.

7.3 Hurricane City Limits. The built-up part of town is still ahead. Cinders and basalt are both well exposed in road cuts. The scarp of the Hurricane Fault is east of town and extends to the north and south, beyond the small volcanoes and farmland.

10.0 Hurricane business district, Junction of Utah State Highway 59 with Utah State Highway 15. Continue toward Zion National Park on Utah State 15. State Highway 59 leads southeast to Colorado City and Pipe Springs National Monument and rejoins U.S. Highway 89 near Fredonia, Arizona.

A short distance east of the junction Utah State Highway 15 turns north and parallels the trace of the Hurricane Fault.

11.2 Bridge over the Virgin River. The gorge at the bridge (fig. 8.4) is cut through basalt flows which have been offset approximately 200 feet by recurrent movement along the Hurricane Fault. The sliver of basalt on the escarpment to the south is part of the same flow series. Older basalts beneath those exposed at the rim of the gorge to the west are offset approximately 2,000 feet, for equivalent remnants occur on top of the Hurricane escarpment to the east. A third older pulse of faulting is suggested by bedrock offset beneath the basalt.

The Hurricane Fault is visible east of the bridge (fig. 8.5), on the north side of the gorge, where Shnabkaib member of the Moenkopi (fig. 8.2) has been lowered on the west against Kaibab Limestone on the east. Monoclinal folding or drag on the fault has reduced the apparent stratigraphic displacement here.

Hot sulphur springs issue from Kaibab Limestone approximately 100 yards east of the Hurricane Fault exposures, in the gorge of the Virgin River.

12.5 Junction of Utah State Highway 17, from the North, with Utah State Highway 15, Turn East on Highway 15, toward Zion National Park. The community of La Verkin is to the west. Approximately 0.3 miles beyond the junction the highway begins to climb the Hurricane escarpment.

13.4 Road cuts through slivers of Kaibab

Figure 8.5. View eastward of the Hurricane escarpment showing offset slivers of basalt on Kaibab Limestone and older rocks along the upper walls of the gorge of the Virgin River. The basalt flow on the shoulders of the escarpment has been offset by the Hurricane Fault and dropped to near road level west of the fault trace.

Limestone and lower Moenkopi Formation in Hurricane fault zone.

14.0 Trace of eastern part of Hurricane fault zone. Lower red and Virgin Limestone Members of the Moenkopi to the west are faulted against Kaibab Limestone on the east. Some of the joints in the Kaibab Limestone show alteration and possibly thus infer connection with the hot springs now present in the Virgin River gorge to the south. To the east the highway is constructed on the lower Moenkopi and upper Kaibab Formations.

15.2 Turn off of dirt road to the south to scenic viewpoint on the rim of Virgin River gorge and the Hurricane escarpment. The road continues eastward at about the base of the Lower Red and top of the tan Timpoweap Members of the Moenkopi Formation.

17.4 A small oil slick is present in basal parts of the Lower Red Member of the Moenkopi Formation in the small gully. The Moenkopi Formation forms striking badlands to the south, east, and north. Some of the irregular topography to the north is produced by landslides in some of the more shaly Moenkopi beds.

17.7 Side road to the north leads to Hurricane Mesa (Pioneer Mesa) (fig. 8.6), where considerable research work was done on testing of ejection seats in jet aircraft. Manikins were ejected from rocket sleds which were "shot" over the edge of the mesa. Hurricane Mesa is capped by Shinarump Formation.

Figure 8.6. View northward to Hurricane Mesa (Pioneer Mesa) capped by resistant Shinarump Sandstone above slopes on various units of the Monekopi Formation. The escarpment exposes virtually the full thickness of the Moenkopi Formation because the highway in the foreground is on upper Kaibab Limestone or the lower Timpoweap Member of the Moenkopi Formation.

18.2 Junction of side road to Kolob Reservoir to the north. High white peaks on the skyline to the north are "beehives" of Navajo Sandstone. Closer to the highway, to the south and north, the resistant limestone is the Virgin Limestone Member of the Moenkopi Formation.

19.4 **Community of Virgin.** Junction of side road to Virgin oil field to the north. The Virgin field is the oldest oil field in Utah and produces from the Timpweap Member at the base of the Moenkopi Formation at an average depth of approximately 550 feet.

20.0 Narrow road and cut where the Virgin River swings in close to the roadway. Virgin Limestone is well exposed at the top of the cuts. Basalt boulders are from flows which cap part of the plateau margin on the north.

21.6 Brick red Middle Red Member of the Moenkopi is exposed. Columnar jointed basalt rims the valley on the north and apparently is part of a series of flows from Crater Hill, a volcano on the flank of the Shinarump-armoured shelf to the north.

To the south, the upper part of the Moenkopi Formation and the overlying Triassic and Jurassic sandstones form needles and pinncales on the skyline. Smithsonian Butte is the westernmost outlier of Navajo Sandstone visible here.

24.1 View northward is over basalt flows to Mt. Kinnesava in Navajo Sandstone on the southwestern end of Navajo Sandstone outcrops south of Zion Canyon. Carmel Limestone forms the flat cap on top of the Navajo beds. Upper Moenkopi, Shinarump, Chinle, Moenave, and Kayenta beds are exposed below the Navajo cliff. To the south, across the river, Middle Red, Shnabkaib, and Upper Red Members of the Moenkopi Formation form striped exposures beneath the Shinarump Sandstone capping Wire Mesa (fig. 8.7).

25.6 Highway is now on Shnabkaib beds beneath the more consistently brick red Upper Red Member and the Shinarump ledge. Eagle Crags to the southeast is strongly-jointed Kayenta and Navajo Sandstone (fig. 8.5).

27.4 Entering Rockville. To the north the strongly-jointed Triassic and Jurrasic sediments form massive cliffs like that characteristic of Zion Canyon.

29.8 Entering Springdale. Shinarump Sandstone is exposed close to the road on the north. Navajo Sandstone forms the cliffs on the skyline. Slope immediately below the Navajo is in Kayenta beds, above the angular red cliff of Springdale Sandstone. Orange red, well-bedded slope formed below is Dinosaur Canyon Member of the Moenave Formation. Chinle beds are covered in the lower slope.

Figure 8.7. Monekopi Formation and overlying Shinarump Sandstone exposed along the north flank of Wire Mesa, as seen southward across the Virgin River from approximately Mile 24.0.

Figure 8.8. View eastward of Eagle Crags, formed in strongly-jointed Wingate and Kayenta Sandstones capping slopes on older Triassic formations below. The prominent ledge at middle height is the Shinarump Sandstone which separates the Moenkopi rocks, below, from the overlying Chinle and Moenave Formations.

31.4 Maroon, gray, and purple Chinle beds exposed in lower slopes of the valley and near the road.

33.1 Boundary Zion National Park. Navajo Sandstone forms the high vertical cliffs at the canyon rim, above a slope carved on Kayenta beds. In the park area Kayenta Sandstone rests on Moenave Formation which here has the resistant cliff-forming Springdale Sandstone above the slope-forming Dinosaur Canyon Member. Chinle beds, below are not well exposed.

34.1 Visitors Center (fig. 8.9)

34.9 Junction of Zion Park Road with Utah State Highway 15, at East End of Bridge Over Virgin River. The loop road leads north along the Virgin River into the scenic center of the park. For a guide along that route see Geologic Guide Segment 8A. Utah Highway 15 leads east up the canyon and out of the park toward Mt. Carmel Junction and U.S. Highway 89.

Figure 8.9. Rocks exposed along the canyon walls in the vicinity of the Visitors Center in Zion National Park. Navajo Sandstone forms the cliffs above a slope on Kayenta Sandstone and upper Moenave Formation forms the lower ledge above the canyon floor.

35.4 Begin climb up switchbacks out of the floor of the tributary, across a small bridge. Contact of the cliff-forming Springdale and lower slope-forming Dinosaur Canyon Members of the Moenave Formation are just above bridge level.

35.9 Hairpin curve on switchbacks is at the approximate top of the Springdale Member of the Moenave and base of the Kayenta Formation. The top of the cliff of Springdale Sandstone shows well on the north wall of the canyon at about this level.

37.0 West Temple in Navajo Sandstone visible to the west across the Virgin River gorge and behind the Visitors Center. On the north the Temple Cap Member of the Navajo Sandstone caps the sheer wall. The road here is in Kayenta Sandstone in the slope below the Navajo Sandstone cliffs.

38.3 West Entrance to Tunnel in Navajo Sandstone. Lookout openings were formerly available for parking but are now blocked by curbing.

39.5 East Entrance to Tunnel along the south wall of Clear Creek. From here to the east the highway climbs up through the upper part of the very spectacularly cross-bedded Navajo Sandstone. Sand for the dunes which formed much of the accumulation was swept in from the northwest, judging from the relatively consistent cross beds dipping toward the southeast. Sand accumulated on the lee of the dunes. A side trail from the parking area leads westward to overlook of the lower Clear Creek and Virgin River Canyons.

40.8 Short tunnel in Navajo Sandstone the road continues along the flanks of Clear Creek gorge in cross-bedded Navajo Sandstone. Stratification is well defined by iron stains which have affected the more porous zones.

44.7 Viewpoint for Checkerboard Mesa to the south in jointed and cross-bedded Navajo Sandstone (fig. 8.11).

45.7 East Entrance to Zion National Park. The road is now rimmed with upper-

Figure 8.10. Cross-bedded Navajo Sandstone exposed on the east entrance road in Zion National Park. The view is toward the northwest from approximately Mile 42.0 and shows the prominent southeastward dipping cross-bed sets in the Navajo Sandstone. The high peak is capped by basal beds of the overlying Carmel Formation.

Figure 8.11. View southward of Checkerboard Mesa near the eastern end of Zion National Park from Mile 44.7. Cross-bedded Navajo Sandstone is broken by nearly vertical joints to produce the peculiar retangular pattern. Cross bedding shows well in the Navajo Sandstone and dips toward the south and southeast. Resistant units at the top of the mesa are lower beds of the Carmel Formation.

most Temple Cap Member of the Navajo Sandstone, with some flat well- bedded basal Carmel Limestone visible here and there.

46.9 Road is now just below Carmel and Navajo contact which is here poorly exposed. The highway continues to climb and next set of roadcuts are in the lower Carmel Formation on the plateau surface on top of the White Cliffs. Cretaceous rocks form Clear Creek Mountain to the north (fig. 8.12).

Figure 8.12. Cretaceous rocks exposed along the south flank of Clearcreek Mountain as seen from approximately Mile 49.0. Tropic Shale forms the slope in the foreground, beneath cliffs of Wahweap-Straight Cliffs and the Kaiparowits Formations.

47.6 Road cuts in basal fossil-bearing part of the Carmel Formation.

49.6 U.S. Highway 15 is still in the lower part of the Carmel Formation, but to the east the barren landslide-scarred hill is in the Tropic or Mancos Shale and underlying Dakota Sandstone (fig. 8.13). In the canyons to the south massive gypsum in the Carmel Formation forms a prominent ledge above the candy-striped lower part of the formation. In the far distance to the east, white Navajo Sandstone holds up the eastward extension of the White Cliff, on the eastern uplifted block of the Sevier Fault.

Figure 8.13. Lower Cretaceous and upper Jurassic rocks exposed east of the highway at Mile 49.6. Slumped Dakota Sandstone and Tropic Shale have interrupted the former route of the highway which diagonaled across the barren landslide mass. Massive gypsum in the upper part of the Carmel Formation forms the ledge in the lower part of the photograph, above the striped lower Carmel Formation which is exposed in the gorge at the extreme lower right.

50.9 View to the southeast in the canyon shows the section from the top of the Navajo Sandstone up through Dakota Sandstone, across the canyon at about road level, overlain by the gray slopes of the Tropic Shale and capped by a basal sandstone of the Straight Cliffs Formation. Hummocky roadway here is a result of heaving and landslide adjustment in the same shaly beds as exposed to the southeast.

51.7 Gypsiferous ashy shale of the coal-bearing Dakota Sandstone is exposed in double road cuts at the crest of the ridge beyond bridge over Meadow Creek. The highway used to pass to the south through the landslide area, but apparently was rerouted to avoid the constant problems produced by the gradual movement.

54.4 Poor exposures of Carmel Formation in road cuts and in gullies to the north and south across the plateau surface.

55.2 Reddish middle part of the Carmel Formation exposed in road cuts. Directly ahead the White Cliffs show interfingering of Carmel and Navajo Formations at the top of sheer cliffs.

56.3 The highway now descends steeply down into the middle red part of the formation. The rim on the north is held up by the massive gypsum of the middle part of the formation. To the north, directly down the road from here, the abrupt westward termination of the White Cliffs shows well where slope-forming gray Cretaceous rocks have been downdropped against the cliff-forming white Navajo Sandstone along the Sevier Fault.

56.8 Cross a small fault which drops the middle redbeds of the Carmel Formation down against the lower limestone beds of the formation (fig. 8.14). *Pentacrinus* is common in the limestone beds east of the fault.

Figure 8.14. Small-displacement fault exposed in the Carmel Formation along the north side of the road at approximately Mile 56.8. The red upper part of the formation, on the left, has been downdropped against the fossiliferous gray limestones and green shales of the lower part of the formation on the right.

Figure 8.15. Carmel Formation exposed north of the highway and west of U.S. Highway 89 at Mt. Carmel Junction. The upper part of the lower ledges are fossiliferous and contain abundant *Pentracrinus* columnals. Other fossils also occur in the slope above.

57.1 Junction of Utah State 15 with U.S. Highway 89 at Mt. Carmel Junction. Abundantly fossiliferous lower Carmel Limestone is exposed in low ledges near the valley floor to the northwest of the junction (fig. 8.15). For a description of the geology along U.S. Highway 89 see Geologic Guide Segment 2.

Segment 8A

0.0 Junction of Road into Zion Canyon National Park Area with Utah State Highway 15. Turn North into Park Area Toward Lodge. The check dam area on the Virgin River above the junction and bridge is one of the local swimming areas. The park road leads northward through debris of the Moenave and overlying formation.

0.8 View up the canyon shows the various formations exposed in the park. They are, from the top down: horizontally-bedded Carmel Formation capping the rim; white cliffs of the upper part of the Navajo Formation. Kayenta Formation forms the slope zone below the Navajo cliffs and overlies the prominent ledge-forming Springdale Member of the Moenave Formation. The lower slope zone is on the Dinosaur Canyon Member of the Moenave Formation. Springdale Sandstone helps to produce rapids a short distance upstream.

1.3 Landslide area makes the narrow V-shaped inner gorge and the debris-littered slopes above. Kayenta beds apparently have slumped to restrict the Virgin River (fig. 8.16). Kayenta Sandstone blocks are strewn over the more shaly active lower part of the formation. The slump or landslide area appears to have formed a dam across the Virgin River in the past and is probably responsible for development of the broad flat bottom of the gorge in the area upstream near the lodge.

Figure 8.16. Slumped Kayenta Formation in the narrows of the Virgin River at Mile 1.3. The landslide mass has produced the rapids.

2.4 Court of the Patriarchs parking area on the east. These almost half-dome appearing erosional remnants of Navajo Sandstone form the western wall of the canyon (fig. 8.17).

3.4 Junction of side road east to the lodge area. Parking area on the west side of the road provides access to trails across the bridge over the Virgin River.

4.2 Campground and picnic area. The canyon bottom is wooded with large Fremont poplar as well as a variety of smaller trees and shrubs. Road cuts on the east side of the canyon just beyond the campground are shaly beds and sandstone in upper Kayenta Formation.

Figure 8.17. View westward of the Court of the Patriarchs as seen from the parking area at Mile 2.4. Massive Navajo Sandstone forms the vertical wall and is capped by thin fossiliferous lower limestones of the Carmel Formation.

5.5 Side road to Weeping Rocks parking area 0.1 mile east of the main park road. Weeping Rocks is a spring area where water that has percolated down through the porous Navajo Sandstone seeps out on top of shales in the Kayenta Formation. Hanging gardens are growing at the contact and have helped develop a protected overhang at the cliff base.

5.8 Parking Area on the South at Great White Throne Turnout. The Great White Throne is the sheer-walled monolith to the southeast (fig. 8.18), with light gray Navajo Sandstone capping the lower reddish part of the formation. Slopes in the immediate vicinity of the parking area and at the base of the Great White Throne are in Kayenta Formation.

Figure 8.18. The Great White Throne as seen southward from the parking area at Mile 5.8. The Great White Throne is composed of Navajo Sandstone capped by basal beds of Carmel Formation. Lowermost slopes in the foreground are on upper units of the Kayenta Formation.

7.3 Parking Area at the End of the Road at the Virgin River Narrows. A trail leads on up the Virgin River Canyon for slightly over 1 mile to where the river occupies nearly the entire width of the gorge. The Narrows of

Figure 8.19 View northward from the parking area at Mile 5.8 toward The Narrows of the Virgin River, the vertical walled gorge in the lower center of the photograph. Navajo Sandstone forms the canyon walls.

the Virgin River are in Navajo Sandstone (fig. 8.19), above where the Kayenta Formation goes below river level. Return down canyon toward the junction with Utah State Highway 15.

8.7 Angels Landing Turnout. The reddish spur and erosional remnant in the foreground south of the river frame the lower part of the Great White Throne.

Segment 9

0.0 Junction of U.S. Highway 89 with Alternate U.S. Highway 89 in Southeastern Kanab. U.S. Highway 89 leads east toward Page. Alternate U.S. Highway 89 leads south to the north rim of Grand Canyon, and rejoins U.S. Highway 89 at Bitter Springs, Arizona, south of Page. U.S. Highway 89 east of town is on the Chinle Formation, with Moenave Sandstone forming the cap of the Vermillion Cliffs to the north (fig. 9.1, 9.2).

1.5 Excellent exposures of the shaly bentonitic lower part of the Chinle Formation on either side and in road cuts near the bend in the highway. View to the south shows the Shinarump Sandstone, forming a prominent cuesta, dipping to the north beneath the subsequent valley formed in Chinle beds. Moenkopi beds beneath the Shinarump ledge forms the Chocolate Cliffs. Kaibab Limestone forms the prominent, almost level skyline to the southeast, south, and southwest on the Kaibab Plateau.

9.5 Side road junction of Utah State Highway 136 up Johnson Canyon to the north. Several movies were made here in the red rocks region in the forties and fifties, when Kodachrome was just coming into it's own. Upper Chinle beds are exposed below Moenave Sandstone on either side of the canyon mouth (fig. 9.3). To the south, the light colored cuesta is capped by Shinarump Conglomerate above the chocolate brown Moenkopi Formation. The highway continues to the east in the middle part of the Chinle Formation.

15.0 Massive Moenave Sandstone continues to form the Vermillion Cliffs to the north and is capped with a thin tongue of white Navajo Sandstone (fig. 9.4). White exposures of sandstone in the junipers to the south are of the upper part of the Shinarump Conglomerate.

16.0 Shinarump Conglomerate (fig. 9.5) is exposed in road cuts on the southwest of the Paunsaugunt Fault which the road crosses here. East of the fault Shinarump Conglomerate caps a low cuesta to the north of the highway above light colored, striped beds of the gypsiferous upper Moenkopi Formation (fig. 9.6). To the southeast, the broad valley of Petrified Hollow is carved on Moenkopi beds with Kaibab Limestone exposed beyond where it is sharply flexed up along the West Kaibab Monocline. The road eastward is in Moenkopi Formation.

19.5 Ripple marked platy sandstone and red gypsiferous mudstone of the Moenkopi Formation exposed in low bluffs both north and south of the road. The lighter more tan units are probably tidal flat-dominated deposits and are slightly more gypsiferous than adjacent beds.

24.8 Crest of small hill. Lower Moen-

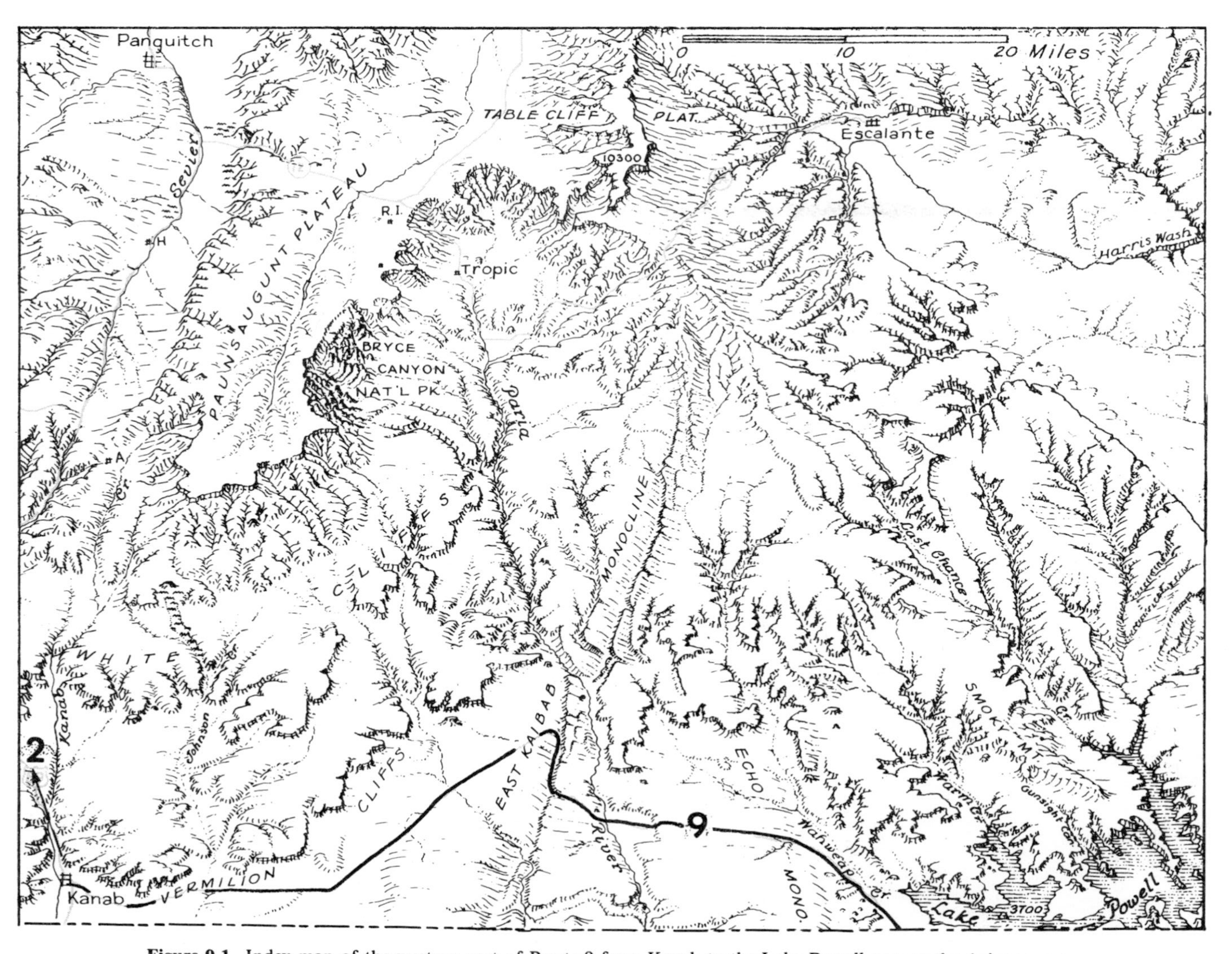

Figure 9.1. Index map of the western part of Route 9 from Kanab to the Lake Powell area at the Arizona state line. Route 2 leads northward from Kanab at the western end of this segment. Route 9 skirts around the Vermillion Cliffs and crosses a series of monoclines before reaching Lake Powell.

Figure 9.2. View northward from Mile 1.0 of Moenave Formation in the cliffs at the east end of Kanab. Prominent sandstone ledges are in the Springdale Member and the reddish banded slope zone below is in the Whitmore Point Member of the formation.

Figure 9.3. View northward of The Grand Staircase from east of Kanab looking northward up Johnson Canyon to the Bryce Canyon National Park area on the skyline. The lowest cuesta escarpment is the Vermillion Cliffs. The next prominent escarpment is the White Cliffs, with the Pink Cliffs capping the skyline in the far distance. Only the backslope of the Chocolate Cliffs cuesta on Moenkopi and Shinarump rocks is exposed in the lower part of the photograph. Moenave Formation holds up the Vermillion Cliffs, Navajo Sandstone the White Cliffs, and the Cedar Breaks or Wasatch Formation holds up the Pink Cliffs. (Photo by W.K. Hamblin)

kopi beds capped by greenish platy sandstone exposed to the north. The juniper and pinion-covered areas to the south are on Kaibab Limestone. Contact of the Moenkopi and Kaibab Formations is a short distance south of the road.

27.6 Cross Buckskin Gulch. Lower Moenkopi beds are exposed in road cuts to the east, but Kaibab Limestone is well exposed in cliffs in the canyon to the south.

32.3 The road continues in the lower part of the Moenkopi Formation. Juniper-covered hills 1,000 yards to the south are on upper beds of Kaibab Limestone. To the north, pinnacles of white, lower Navajo Sandstone rise above the prominent Vermillion Cliffs of the Moenave Formation. The slope zone between here and base of the cliffs is on the Chinle Formation and the upper part of the Moenkopi Formation.

34.0 Side Road to Old Paria Townsite (fig. 9.7). A primitive road leads approximately 4 miles to the Paria townsite through spectacularly colored Triassic rocks. East of here upper Kaibab Limestone is exposed on the crest of the East Kaibab Monocline where the beds are almost flat. A short distance farther east, however, Kaibab rocks bend down abruptly on the monocline. It is the monocline flexure, in part, which has helped produce the striking scenery in the vicinity of Paria, where thick, brightly colored Moenkopi and Chinle beds are spectacularly exposed.

35.2 Near the top of Kaibab Limestone at the major bend in the highway built on the east limb of the monocline (fig. 9.8). Moenkopi beds show well on the west flank of the gully, but Chinle rocks are almost totally buried by slumped debris from the overlying Moenave Formation and by recent sand dunes. The road drops southward into a subsequent valley carved in Moenkopi and Chinle beds, above and east of the Kiabab Limestone exposures along the monocline.

BRYCE CANYON - KANAB AREA

System	Group/Formation	Unit	Thickness	Notes
TRIASSIC	Moenave Fm	Springdale Ss M	60	
		Whitmore Pt M	0-50	*Seminotus (fish)*
		Dinosaur Can M	80-140	
	Chinle Fm	Petrified Forest M	400	
		Shinarump M	0-160	
	Moenkopi Fm	upper red Shnabkaib Member - 200' middle red Virgin Limestone Member - 50' lower red	700—2000	bedded gypsum *Tirolites*
PERMIAN		Kaibab Ls	250	Paleozoic beds known from well data
		White Rim? Ss	200	
		Toroweap Fm	100—500	
		Hermit Shale	0-100	Organ Rock equiv.
		Queantoweap Sandstone	1000	Cedar Mesa equiv
PENN		Callville Ls	200—600	Hermosa equivalent
MISS	Redwall Ls	Horseshoe Mesa M	170-290	cherty
		Mooney Falls M	240	
		Thunder Spr M	210	
		Whitmore Wash	100	cherty
D		Ouray Limestone M	0-150	
		Temple Butte Ls	110-250	Elbert equivalent
CAMBRIAN		Muav Limestone	1120—1230	
		Bright Angel Sh	330	
		Tapeats Ss	290—330	
LATE P-€		Grand Canyon Series	820+	

System	Group/Formation	Unit	Thickness	Notes
Q		basalt flows	0-300	locally present
P		Sevier River Fm	0-60	
EO - OLIG - MIO		Volcanic rocks thickens from Panguitch area northword toward Marysvale area	0—several hundred	tuffaceaus andesites
		Brian Head Fm	0-300	
PALEO		Cedar Breaks Fm (=Wasatch Fm)	600	red and varicolored Bryce Cyn beds, fresh water deposit
		Pine Hollow Fm	0-400	
CRETACEOUS		Canaan Peak Fm	0-1000	
		Kaiparowits Formation	265—700	mostly bluish gray silts and sands, *fresh water fossils*
		Wahweap and Straight Cliffs Sandstones, undivided	535—1620	
		Tropic Shale = Mancos	625	marine beds *Inoceramus* *Sciponoceras*
		Dakota Ss	175	coal beds
JURASSIC		Henrieville Sandstone	0-300	cut out by pre-Dakota erosion to the west
	Entrada Fm	Escalante Mbr	0-50	
		Cannonville M	0-230	
		Gunsight Butte	0-220	gypsum = "Curtis" of older reports
	Carmel Fm	Wiggler Wash M	0-60	
		Winsor Mbr	180-250	
		Paria Riv Gyp Mbr	60-160	*Pentacrinus abundant*
		Crystal Creek Mbr	0-180	
		Kolob Ls Mbr	70-230	
		Navajo Sandstone	1700—2000	
TRIASSIC		Tenney Canyon Ton. of Kayenta Ss	0-120	
		Lamb Point Tongue of Navajo Ss	400	
		Kayenta Fm (main body)	300	

Figure 9.4. Stratigraphic section of rocks exposed in southern Utah in the Kanab and Bryce Canyon Areas (from Hintze, 1973).

Figure 9.5. Shinarump Sandstone in a small gorge south of the highway at Mile 16.0, in the downdropped block west of the Paunsagunt Fault. Kaibab Limestone is exposed on the East Kaibab Plateau on the skyline in the background.

Figure 9.6. View northward of the Grand Staircase from approximately Mile 15.0 showing minor offsets in the Moenave Formation along the small gully where the Paunsagunt Fault cuts the Vermillion Cliffs. Banded Moenkopi Formation, capped by Shinarump Sandstone, is exposed in the lower right of the photograph. The White Cliffs of Navajo Sandstone rise above the Vermillion Cliffs and the shelf carved on Kayenta Formation. The Paunsagunt Plateau in the background is capped by the Pink Cliffs, the erosional edges of the Cedar Breaks or Wasatch Formation.

Figure 9.7. Paria town site along the Paria River as seen from along the side road which leads north from Mile 34.0. Shinarump Sandstone forms the resistant unit in the foreground and Chinle beds form the striped exposures in the lower part of the Paria Canyon wall. The town site is near the cluster of trees along the river in the center of the photograph. Wingate and Moenave Formations cap the cuesta on the skyline.

Figure 9.8. Steeply dipping upper beds of the Kaibab Limestone on the east side of the East Kaibab Monocline at Mile 35.2. The road in the foreground is on basal beds of the Moenkopi Formation in a strike valley parallel to the monocline.

39.2 The road bends away from the monocline and cuts through The Reef on the east side of the valley and enters double road cuts through Moenave beds which are reddish sandstone units with interbedded mudstone, and then through Kayenta and Navajo beds in cuts ahead.

40.4 Top of the Navajo Sandstone. View toward the north at the east end of the cut is of spectacularly banded Carmel and Entrada Formations on the east flank of the East Kaibab Monocline and are capped by a thin Morrison Formation and Dakota Sandstone (fig. 9.9). The skyline to the northeast is of the Kaiparowits Plateau with Straight Cliffs and Wahweap coal-bearing formations high on the bluff.

Figure 9.9. View northward along the East Kaibab Monocline from Mile 40.4. Uppermost beds of the Navajo Sandstone are the light-colored rocks at the extreme left. Banded valley-forming rocks are in the Carmel Formation and are overlain by ledge-forming members of the Entrada Formation. The tableland at the extreme right is capped by Morrison beds.

42.1 Massive cross-bedded white Gunsight Butte Member of the Entrada Sandstone (fig. 9.4) forms bold exposures in hills to the south.

43.7 **Cross the Paria River,** here dry some of the time. East of the crossing is the junction with the road south to Paria Canyon primitive area. Entrada rocks show well in ledges to the north (figs. 9.10, 9.11). To the south, massive white Navajo Sandstone forms cliffs down Paria Canyon.

46.0 Major bend in the highway. Broad fluvial channel fill is well exposed in the middle of the candy-striped Cannonville Member to the east. The road swings back northeastward to climb up over the white sandstone of the Escalante Member within the Entrada Formation.

Figure 9.10 View northward from east of the Paria River at approximately Mile 44.0 of the ledge-forming Entrada Formation. The lower banded unit is the Cannonville Member and the upper white sandstone cliff-former is the Escalante Member of the formation.

47.5 Drainage divide. To the north and east stretches the Cretaceous coal-bearing Wahweap-Straight Cliffs Formations in the Kaiparowits Plateau (fig. 9.13). These coal-bearing Cretaceous Formations rise in steps above the silvery gray slope carved on the Mancos or Tropic Shale which forms the prominent valleys at the cliff base (fig. 9.12).

50.1 Road cuts through the upper part of the massive white Escalante Sandstone. Striped Carmel beds are exposed to the south. To the north, a simple anticline is expressed at the north end of the Echo Mono-

Figure 9.11. Entrada Formation exposed northward from approximately Mile 46. The well-bedded Cannonville Member of the formation, in the foreground, is overlain by the massive cross-bedded sandstone of the Escalante Member.

Figure 9.12. View northward across Wahweap Canyon to Cretaceous rocks exposed along the south side of the Kaiparowits Plateau. Mancos Shale forms the flat valley bottom and the gray shale slopes below cliffs of the Straight Cliffs Formation, the coal-bearing unit in the Cretaceous section. Upper beds of Entrada Sandstone and a thin Dakota Formation are exposed in the foreground and along the rim of the shallow inner gorge of Wahweap Canyon.

cline by dips and rise and fall of the massive white Escalante Member of the Entrada Formation below the more resistant Dakota Sandstone and. slopes of Mancos Shale.

53.9 Massive white sandstone of the Escalante Member exposed to the north in the core of the anticline.

55.8 Outliers of somewhat cavernous-appearing, cross-bedded Entrada Sandstone on the south side of the road. Silvery gray Mancos Shale forms the prominent slope zone beneath the striped coal-bearing Cretaceous units to the north and northeast, across Wahweap Canyon. From here, the road dips down the east flank of a small anticline on the Echo Monocline essentially on the top of the Entrada Formation.

57.8 Side Road. Entrance to Glen Canyon City on the north. Beyond the development Dakota Formation is well exposed in the canyon to the north, beneath the silvery striped Mancos Shale.

60.2 White massive cross-bedded sandstone of the Entrada Formation. This same unit is exposed along Wahweap Creek to the north (fig. 9.14) and is capped by massive brown sandstone bluffs of the Dakota Formation. Above that, the Mancos Shale forms a typical gray slope leading up to the coal-bearing Straight Cliffs and overlying Wahweap Formations on the skyline. Directly ahead, triple stacks of the Navajo Power Plant rise above Page, Arizona. The plant burns coal mined in the Black Mesa area to the east.

61.5 White cross-bedded sandstone at the base of Blue Hole Wash Mesa is Cannonville Member of Entrada, overlain by Escalante Member of the Entrada which forms the prominent, nearly vertical cliff beneath the Morrison-Dakota ledge at the top. Cross Blue Hole Wash beyond the mesa.

62.7 View off to the east of the blue water of Lake Powell. Navajo Mountain

GLEN CANYON DAM - KAIPAROWITS - CIRCLE CLIFFS

System	Group / Formation	Member / Unit	Thickness (ft)	Remarks
TRIASSIC	Glen Canyon Group (cont'd)	Navajo Sandstone	1700	*tracks of 3-toed dinosaur 930' below top at Glen Cyn Dam east diversion tunnel*
TRIASSIC	Glen Canyon Group (cont'd)	Kayenta Fm		
TRIASSIC	Glen Canyon Group (cont'd)	Wingate Ss		
TRIASSIC	Chinle Fm	Church Rock M	0-150	
TRIASSIC	Chinle Fm	Owl Rock Memb	150-250	
TRIASSIC	Chinle Fm	Petrified Forest M	250—400	varicolored bentonitic clays
TRIASSIC	Chinle Fm	Monitor Butte M	100-250	mudstone
TRIASSIC	Chinle Fm	Shinarump M	0-200	ss, cg, mudstone
TRIASSIC	Moenkopi Fm		300—400	
PERMIAN	Kaibab Ls		0-200	Paleozoic rocks known from well data
PERMIAN	Cutler Group	White Rim Ss	400	
PERMIAN	Cutler Group	Toroweap Fm	0-100	
PERMIAN	Cutler Group	Organ Rock Sh	300—500	
PERMIAN	Cutler Group	Cedar Mesa Sandstone	1200—1300	Queantoweap equiv
PENN	Hermosa Gp	Honaker Trail Fm	600	Callville equivalent
PENN	Hermosa Gp	Pinkerton Trail Fm	150-200	
MISS	Redwall Ls	Horseshoe Mesa M	175	cherty
MISS	Redwall Ls	Mooney Falls M	200-300	
MISS	Redwall Ls	Thunder Spr M	150-200	
MISS	Redwall Ls	WhitmoreWash M	100	
DEV	Ouray Ls		115-140	
DEV	Elbert Fm		150-250	Temple Butte Ls equivalent
CAMBRIAN	Muav Limestone		630	
CAMBRIAN	Bright Angel Sh		300	
CAMBRIAN	Tapeats Ss		200	
P€			—	not yet penetrated

System	Group / Formation	Member / Unit	Thickness (ft)	Remarks
T	Cedar Breaks Fm		0-200	*fresh water snails (Schneider, 1967)*
CRETACEOUS	Kaiparowits Formation		2700	predominantly fluvial; bluish gray siltstone and sandy siltstone; *fresh water clams*; *duckbill dinosaur bone*
CRETACEOUS	Wahweap Sandstone		1250	near shore and marine deposits
CRETACEOUS	Straight Cliffs Fm	Drip Tank Mbr	140	
CRETACEOUS	Straight Cliffs Fm	John Henry Mbr	740	coal
CRETACEOUS	Straight Cliffs Fm	Smoky Hollow M	115	coal
CRETACEOUS	Straight Cliffs Fm	Tibbet Cyn Mbr	105	coal
CRETACEOUS	Tropic Shale		610—640	*Inoceramus*
CRETACEOUS	Dakota Ss		40-100	UNCONFORMITY
JURASSIC	Morrison Fm	Salt Wash Memb	0-365	
JURASSIC	Cow Springs? Ss		0-145	white sandstone
JURASSIC	Entrada Ss		600—750	
JURASSIC	Carmel Fm		200—400	
JURASSIC	Glen Canyon Group (cont'd)	Navajo Ss	(1700)	

Figure 9.13. Stratigraphic section of rocks exposed in the Kaiparowits Plateau, Circle Cliffs, and Glen Canyon Dam area in southern Utah (from Hintze, 1973).

Figure 9.14. View northward across entrenched Wahweap Canyon from approximately Mile 60.2. Massive cross-bedded upper Entrada Sandstone is exposed in the gorge and is capped by a thin, resistant, ledge-forming unit of the Morrison Formation. Higher slopes beyond are in Mancos Shale and are capped by coal-bearing Straight Cliffs Formation.

Figure 9.15. View northward from approximately Mile 68.5 to the Wahweap Marina and the arm of Lake Powell that extends up Wahweap Creek. Various units of the Entrada Formation form cliff exposures beyond Lake Powell.

rises on the far skyline due east. Escalante and Cannonville members of the Entrada Formation show well on the south side of the road with the massive, more platy bedded upper Escalante beds now starting to take on a few reddish streaks above the softer cross-bedded, silty Cannonville rocks.

65.5 Utah-Arizona State Line. A short distance beyond is a side road into trailer court and camping area.

68.5 Side Road to Wahweap Marina and other facilities on the west side of Lake Powell. The marina can be seen to the northeast along the lake shore (fig. 9.15).

70.6 Highway drops steeply down into the Glen Canyon Area, in Navajo Sandstone. The Carmel Formation is a thin red rippled sandstone unit underlying much of the Page townsite and capping the bluff between here and there.

71.5 View of the east abutment of the dam in cross-bedded Navajo Sandstone well exposed in bluffs along the immediate side of the road, as well as down in the gorge, near the dam area (fig. 9.16).

Figure 9.16. Glen Canyon Dam as seen north up the gorge of Glen Canyon from the overlook west of Page. Navajo Sandstone makes up most of the vertical walls of the gorge.

72.5 Double road cuts through upper beds of Navajo Sandstone. The typical wind-blown cross bedding of the unit shows well here and in the overlying beds as well.

72.8 Side Road to the Glen Canyon Visitors Center, and west abutment to the bridge over Glen Canyon. Glen Canyon Dam ponds Colorado River water in Lake Powell and is a concrete structure based in large part in Navajo Sandstone.

74.0 Junction Loop Business U.S. Highway 89 into Page. Turn east on business route and climb bluff into Page.

74.8 Entering Page. Bluffs in the immediate vicinity of Page community are capped by Carmel Formation above the Navajo Sandstone. Continue on business route U.S. 89 through town.

76.4 Junction of Business Route U.S. Highway 89 with U.S. Highway 89 Bypass. Turn south toward Cameron, Grand Canyon, and Flagstaff. Deeply incised Glen Canyon is visible occasionally to the west. The road continues for the next several miles through the uppermost beds of Navajo Sandstone veneered here and there by recent windblown sand.

79.3 Cross beneath power transmission lines heading toward the south. The "P" on the bluff to the east is on upper beds of Navajo Sandstone.

82.2 Double road cuts through Navajo Sandstone. The high skyline ahead to the south-southeast is the top of the Navajo Sandstone capped by basal beds of the overlying Carmel. The Carmel Limestone has been stripped off from near the road and has left a planar surface into which the small tributaries of Glen Canyon have now carved deep gorges.

86.7 Road bend, excellent view back to the north of the panorama of the basin to Smoky Mountain and the Cretaceous rocks beyond Lake Powell. Navajo Sandstone is well exposed in deep double road cuts that mark the backside of the cuesta.

88.3 Road cuts at the top of the divide in Navajo Sandstone. The pasturelike upland surface here has rolling relief and suggests recent removal of the overlying Carmel rocks.

97.5 Beginning of Deep Roadcuts through the deep red Navajo Sandstone at the top of Echo Cliffs. **Rest area at the West End of the cuts looks over the Marble Platform,** which is developed on the top of the Kaibab Limestone (fig. 9.17). Across the Marble Platform and the Marble Gorge of the Colorado River to the northwest, the Vermilion Cliffs of Moenave and Navajo

Figure 9.17. View westward across the Marble Platform from the rest area at the west end of the cuts through the Echo Cliffs at Mile 97.5. The Colorado River is entrenched here into a surface held up by resistant Kaibab Limestone. The Vermilion Cliffs of Moenave and Navajo Formations rise above a slope zone eroded on well-bedded Moenkopi and soft Chinle beds.

Sandstones form the bold escarpment. The cliffs loop around the east side of the platform, and then heads south along the Echo Cliffs Monocline. Moenkopi rocks form the Chocolate Cliffs in the reentrant to the north below the Vermilion Cliffs. South of the rest area, Navajo Sandstone caps the cliff well above the Moenave beds (fig. 9.18). Alternate U.S. Highway 89 is the road in the subsequent valley to the southwest and is built on Moenkopi beds above eastward dipping Kaibab Limestone.

Figure 9.18. View southward along the Echo Cliffs and the Echo Cliffs Monocline from approximately Mile 98.0. The cliff is capped by Navajo Sandstone and the bedded ledge and slope zone below is on the Moenave Formation which rises above the flat land on Chinle and Monekopi beds. Kaibab Limestone is exposed in the Marble Platform in the distance at the right.

The highway drops down the cuesta face through loose slumped material. Chinle Shale beneath is unstable when wet, and the overlying Moenave rocks have been jointed and slumped. Such movement produces angular rubble blocks along the road as well as the hummocky topography to the west.

99.2 Slumped pink and gray ashey-appearing Chinle Shale in the upper part of cuesta is veneered by an armor of slumped Moenave, Kayenta, and Navajo Sandstone blocks.

101.3 Junction of Alternate U.S. Highway 89A with Main U.S. Highway 89 at Bitter Springs. Turn south toward Flagstaff on U.S. Highway 89. To the west of the junction, Kaibab Limestone is exposed in the gullies. Upper Chinle, Moenave, and Navajo formations form the Echo Cliffs to the east. The highway continues to the south toward Cameron in a subsequent valley carved on the Moenkopi and lower Chinle formations.

103.0 Shinarump Conglomerate and tilted slumped toreva blocks of overlying Chinle beds exposed east of the road at the base of the escarpment and along Roundy Creek. Top of the Kaibab Limestone is well exposed in canyons and gullies to the west.

106.1 Kaibab Limestone exposed in cuts immediately west of the road. The road continues about on the top of the Kaibab.

110.2 U.S. Highway 89 leaves upper beds of the Kaibab Limestone and diagonals across the Moenkopi Formation which is a thin-bedded, chocolate-colored, silty, mudstone. Moenkopi exposures are capped by tan resistant Shinarump Conglomerate.

112.3 Side roads leading up onto the Marble Platform from Roundy Creek show Moenkopi rocks beneath a thin Shinarump Conglomerate. Kaibab Limestone is exposed in the headwaters of the gullies to the west. On the Echo Cliffs to the east, the light-colored striped tan and reddish brown sandstone is the Navajo Formation resting on the darker reddish brown Moenave Formation. Most of the valley between the road and the base of the cliff is carved on the Chinle Formation (fig. 9.19).

120.4 Trading Post at Cedar Ridge. The road continues in red brown Moenkopi beds with Kaibab beds exposed along the monocline to the west and Chinle and younger formations making the Echo Cliffs to the east. The younger rocks are closely jointed and break down to produce gaps in the cliffs, allowing access to the plateau surface to the east.

122.3 Double road cuts through Moenkopi beds. Kaibab Limestone exposed a short distance to the west of the road, and Shinarump Conglomerate forming the light-colored sandstone cuesta-cap to the east in the headwaters of Hamblin Wash (fig. 9.20).

125.8 Moenkopi and Shinarump Formations exposed in road cuts and bridge abutment on the southwest side of the road.

127.8 The Gap Trading Post with school, and other facilities. Upper Chinle

Figure 9.19. View northeastward to the face of the Echo Cliffs at approxinately Mile 112.0. The slope at the base of the escarpment is on Chinle beds with the lowermost persistant ledge held up by the Dinosaur Canyon member of the Moenave Formation. The overlying slope is carved on the Whitmore Point Member and the next dark banded series of the ledges are in the Springdale Sandstone Member of the fornation. Navajo Sandstone caps the escarpment and forms the upper desert varnish-painted sheer wall.

Figure 9.20 View northeastward from the top of the Kaibab Limestone to jointed Navajo Sandstone along the Echo Cliffs Monocline at approximately Mile 122.0. Banded exposures in the gorge in the foreground are Moenkopi Formation overlain by Shinarump Sandstone. The valley beyond is carved, in large part, in the Chinle Formation.

beds exposed well to the east beneath the angular weathering Moenave Sandstone which forms a prominent cuesta beneath the Kayenta and Navajo Sandstones.

132.7 Chinle beds exposed beneath terrace cover. Here they demonstrate that they are particularly susceptible to heaving, and make poor roadbeds.

136.9 Roadside Rest Area. The road continues through the lower Chinle Formation with Moenkopi rocks exposed to the west. Petrified Forest Member of the Chinle is exposed to the east beneath the more evenly-bedded, lacustrian-appearing pinkish Owl Rock and Church Rock Members of the formation.

137.9 Excellent exposures of Shinarump Conglomerate to the west and near the road with some exposures of Moenkopi beds showing in the farther reaches of the canyon. Outcrop of the overlying Moenave Sandstone now begins to swing to the southeast around the southwestern edge of the San Juan Basin.

138.5 Junction U.S. Highway 160 with U.S. Highway 89. U.S. Highway 89 continues south to Cameron and Flagstaff and the entrance road to Grand Canyon National Park. U.S. Highway 160 leads northeastward across the Navajo Reservation to the Four-Corners area. For a description of the geology along U.S. Highway 160 see Guide Segment 10.

Segment 10

0.0 Junction of U.S. Highways 160 and 89. Segment 10 follows U.S. Highway 160 east toward Tuba city and Kayenta. U.S. Highway 89 leads south toward the entrance to Grand Canyon National Park and Flagstaff. For a route description along U.S. Highway 89 northward from here see Geologic Guide Segment 2.

The road junction is in the Petrified Forest Member of the Chinle Formation. The member is composed of interbedded stream-channel sandstone and varicolored shale and mudstone. This member erodes moderately easily and forms the strike valley to the north and south. From here the route of this guide leads upsection into younger and younger beds of the Chinle Formation.

0.7 Cross Hamblin Wash and rise from the Petrified Forest Member into the pinkish banded Owl Rock Member of the Chinle Formation. The upper member forms pronounced laminated pinkish gray and green badlands, distinctly unlike the rounded Painted Desert-type massive badlands of the underlying member.

1.6 Road rises up through the upper part of the Chinle Formation, a typical wavy to hummocky road. Highway construction is easy across the slope-forming parts of the formation, but holding the road after construction is difficult because the soft volcanic ash-bearing shales heave under load or after wetting and drying. Ashy gray marls and pinkish mudstones show well in badlands along the road (fig. 10.1).

Figure 10.1. View northward to exposures of the Owl Rock Member of the Chinle Formation along the Echo Cliffs Monocline, where Route 10 crosses the main part of the fold.

To the north the monocline along the Echo Cliffs separates the Marble Platform on the west from the Kaibito Plateau on the east. Hamblin Wash has eroded along the outcrop of the Chinle and Moenkopi Formations between the more resistant Kaibab Limestone below and the cliff-forming Moenave and Navajo Sandstones above.

4.8 Side road north to Moenave. Approximately 500 yards north of the highway, along the Moenave Road, three-toed dinosaur foot prints (fig. 10.2) are preserved in one of the resistant sandstone beds of the upper part of the Chinle Formation. The tracks are in a barren area immediately west of the road. Local women may have exhibits of juniper seed and glass beadwork for sale at the site.

Figure 10.2. Three-toed dinosaur footprints in the upper part of the Chinle Formation north of the route on the Moenave road.

Moenave is the community to the north near the tall poplar trees and spring areas along the Moenave Sandstone escarpment.

6.4 U.S. Highway 160 climbs through double road cuts in the almost Moenkopi-looking upper part of the Chinle Formation in the candy-striped gypsiferous upper part of the section. Moenave beds hold up the escarpment to the north (fig. 10.3).

9.6 The highway now swings northward up off the floodplain of Moenkopi Wash

Figure 10.3. View northward to uppermost Chinle beds and lower Moenave beds along the walls of Moenkopi Wash southwest of Tuba City at approximately Mile 8.

through road cuts of Moenave Formation. These resistant sandstone beds form the plateau surface beneath the south end of Tuba City.

10.4 Junction of Access Road North to Tuba City and South to Moenkopi with U.S. Highway 160.

11.7 Small reservoir east of Tuba City. Cross-bedded Kayenta Sandstone rests on top of the resistant Moenave Sandstone near the road. The road east of Tuba City is in poor exposures of Wingate-Kayenta-Navajo Sandstone blanketed by loose drifting sand. Castle Butte to the north and Middle Mesa to the northeast show Navajo Sandstone capped by Carmel Formation.

16.4 Now abandoned uranium processing plant and associated housing. Rounded Navajo Sandstone begins to show in exposures in Moenkopi Wash to the south. Brownish beds in the immediate vicinity are probably basal Carmel Formation.

25.6 Rest area at the western base of Middle Mesa (fig. 10.4). Navajo Sandstone forms the lower, light tan, rounded bluffs and are capped by flat-bedded reddish Car-

Figure 10.4. View eastward from near the rest area at the western base of Middle Mesa at Mile 25.6. Navajo Sandstone is exposed beyond the buildings, below well-bedded Carmel and Entrada Formations.

mel Formation which is overlain by remnants of Entrada Sandstone at the top of the mesa.

33.3 Side Road to Tonalea School. A short distance beyond the junction the road dips down into a wash which exposes upper Navajo Sandstone, crinkly reddish Carmel Formation, and reddish brown to grayish green Entrada Formation. Red Lake Trading Post is on the western bluff of the wash south of the highway. Reddish beds of the Entrada Formation have a mottled to marbled weathered surface. Red Lake is in the valley to the south.

34.8 Elephant's Feet Rest Area. The "feet" are erosional remnants of light gray cross-bedded Entrada Sandstone (fig. 10.5). The "toes" are in the underlying reddish part of the formation. To the east Morrison Formation and Dakota Sandstone are exposed along the western end of Black Mesa.

39.7 Junction of side road to Navajo National Monument, Inscription House, and Navajo Mountain. U.S. Highway 160 swings parallel to the electric railroad right-

Figure 10.5. The Elephants Feet at Mile 34.8. These interesting erosional remnants are in Entrada Sandstone.

of-way. The railroad hauls coal from Black Mesa to the Navajo Power Plant near Page, Arizona.

43.2 Cow Springs Trading Post. The prominent white sandstone to the south of the trading post is the Cow Springs Sandstone (fig. 10.6) and is above the reddish part of the Entrada Formation and below the

Figure 10.6. Cowspring Sandstone is the massive light gray unit at the base of the escarpment on the north side of Black Mesa. These exposures are southeast of Cowspring Trading Post, south from approximately Mile 44.0.

Morrison Formation. Prominent gray shale slopes near the top of the mesa expose Mancos Shale capped by the coal-bearing Cretaceous Mesa Verde Group.

47.8 Coconino County—Navajo County Boundary. The Cow Springs Sandstone forms the relatively prominent light gray ledges to the south beneath Morrison beds. Morrison Formation forms the lower half of the Black Mesa escarpment and Mancos Shale and Mesa Verde beds the upper half. From high points along the road Navajo Mountain is visible toward the north. Navajo Mountain is cored by a laccolith, like intrusions in the Henry Mountains, La Sal Mountains, and Ute Mountain, but Navajo Mountain is still roofed with uparched Navajo Sandstone.

51.4 Junction of Arizona State Highway 98 with U.S. Highway 160. Arizona Highway 98 leads northwestward toward Page. To the south the front of Black Mesa still exposes Morrison beds as the gray and pinkish units near the base, with Mesa Verde Sandstone forming the mesa cap. U.S. Highway 160 continues on through Klethla Valley which is carved on nonresistant Carmel and Entrada beds.

63.6 Storage Silos and Tipple at the End of the Railroad. Coal is transported from the mines on the mesa to the southeast to the loading area by a belt system which is bridged over the highway (fig. 10.7). Coal in the strip mines area is up to 36 feet thick and as much as 100 feet of overburden is being removed in the mining operations (fig. 10.8).

64.4 Road Junction to Navajo National Monument Toward the North. A side road also leads south from here approximately 15 miles into the Peabody Coal Mine area on top of Black Mesa. Morrison and overlying Cretaceous rocks are exposed along the access road in the mesa front (fig. 10.7).

East of the junction there is a double series of cliffs on the escarpment. Lower ones are held up by Morrison Formation above the light gray Cow Springs Sandstone and the upper one is held up by sandstone of the Mesa Verde Group, above the gray Mancos Shale slope.

Figure 10.7. Eastward along the escarpment of Black Mesa from near Mile 64.4, past the belt delivery system for Peabody Coal Company's operation. the belt delivers coal to storage and loading facilities at the end of the Navajo project railroad.

Figure 10.8 Peabody Coal Company Mine in Cretaceous Mesa Verde rocks on Black Mesa south of the route. Approximately 100 feet of overburden is being stripped from left and heaped in chat piles on the right to gain access to underlying coal that is up to 36 feet thick.

66.2 Rest Area. Massive very light gray Cow Spring Sandstone is exposed at the base of the mesa scarp and is overlain by reddish shale and light cliff-forming sandstone of the Morrison Formation. Behind the frontal lower cliff zone of Morrison Formation (fig. 10.9) Mesa Verde rocks are exposed in the prominent banded cliffs on the skyline. Toward the north Navajo Sandstone and older units are exposed on the south end of Organ Rock Monocline at the northern edge of the Black Mesa Basin.

Figure 10.9. View eastward of the very sandy lower part of the Morrison Formation exposed in a frontal cuesta along the north flank of Black Mesa basin at approximately Mile 66.2. The thick sandstones are channel filling and are separated by slopes carved on gray green shale.

69.3 Massive slickrock exposures to the north are the top of the Navajo Sandstone dipping southward on the monocline. Klethla Valley through here is a subsequent valley in Carmel and Entrada beds.

72.6 Tsegi Trading Post. The highway is constructed across fill and debris from the escarpment to the south and across soft Entrada and Carmel beds. Massive Navajo Sandstone, slope-forming Kayenta Sandstone, and cliff-forming Wingate Sandstone are visible in canyons to the northwest (fig. 10.10) where erosion has cut down into the still older Chinle beds along the Organ Rock Monocline in Tsegi Canyon.

73.5 Road cuts are in upper Navajo Sandstone along the narrows of Laguna Creek. Notice how deeply Laguna Creek has entrenched into the Recent sand and silt which partially filled the gorge.

Figure 10.10. View northward up Tsegi Canyon, west of the Tsegi Trading Post at Mile 72.6. Cliffs on the skyline and in the foreground are in Wingate Sandstone. The massive sandstone rests on slope-forming Chinle beds that are exposed in the notch in the canyon at the lower left.

76.2 The Organ Rock Monocline now swings to the north away from the highway. Road cuts are in Morrison Formation. To the north Navajo, Kayenta, and Wingate Sandstones form spectacular flatirons on the monocline. To the south the Morrison Formation is now composed of massive series of channel-fill sandstone lenses. It is these kinds of deposits which have produced dinosaurs in various parts of the West. Uranium minerals accumulations have also been recognized. Cretaceous rocks form the rim beyond on the skyline (fig. 10.11) around Black Mesa Basin.

Figure 10.11. Southeastward from near Mile 80 along the northeastern side of Black Mesa basin showing Cretaceous rocks which hold up the rim. Morrison beds form the lower exposures and Mancos Shale erodes to the distinct slope zone up the middle part of the cliff. Mesa Verde Group forms the upper series of the cliffs along the skyline.

80.0 Double road cuts through sandstone in the Morrison Formation. Numerous small channels are cut into the sandstone, some of them are partially filled with green and red mudstones and others are filled with coarse-grained sandstone. Toward the north Agathlan or Agathle's Needle rises above the cuesta of Navajo Sandstone and older rocks beyond the town of Kayenta.

84.2 Junction of U.S. Highways 160 and 163. Turn north on U.S. Highway 163 toward downtown Kayenta, Monument Valley, and Mexican Hat. U.S. Highway 160 continues to the northeast through the Four Corners region and toward Cortez, Colorado. Numerous small intrusive masses are visible to the east and have the same general appearance as Shiprock or Agathle's Needle, but are smaller.

86.4 Northeast Edge of Kayenta. Cross-bedded sandstone nearby is upper Navajo Sandstone. To the east rises Church Rock, a basaltic volcanic neck. The route soon crosses Laguna Creek which is deeply entrenched into young valley fill. Navajo Sandstone is exposed to the north, dipping southward into Black Mesa Basin.

88.2 Cross through the cuesta composed of Wingate, Kayenta, and Navajo Sandstone, all dipping southward off the Monument Valley upward and into Black Mesa Basin. North of the cuesta the highway is constructed over varicolored purple, green, maroon, and gray Chinle Formation. Chaistla Butte is the dark promontory to the east and is one of the volcanic necks or intrusions associated with Agathle's Needle, the peak to the north (fig. 10.12).

Figure 10.12. Agathle's Needle (Agathlan or El Capitan) is the now-denuded volcanic neck of dark breccia, here rising above gentle slopes on the Chinle Formation. Agathle's Needle is one of a series of similar volcanic necks or intrusions that can be seen east of the highway north of Kayenta.

93.5 Rest Area at the Northwest Base of Agathle's Needle or Agathlan. The spire is a now denuded intrusion which may have been a volcanic neck or a fingerlike plug up into the Triassic Rocks. It is a dark basaltic-looking breccia cut by smaller dikes. Chinle Formation is exposed in the low country side, and here and there gravel caps contain silicified fossil wood. Owl Rock to the west is an outlier of Navajo and Kayenta Sandstone, above the massive Wingate Sandstone.

96.9 Side road to Hoskinnini Mesa. Boot Mesa is to the northwest and is rimmed by a vertical wall of Wingate Sandstone, capped by Kayenta and Navajo Sandstones. The highway is in Chinle Formation.

98.9 Massive Shinarump Sandstone forms the ledges and cliffs to the east and for a short distance to the northeast. Another small intrusion cuts the Shinarump beds to the northeast. The highway crosses a thinned Shinarump Sandstone cuesta at approximately Mile 98.4. Beyond that the road is on the Hoskinnini Member of the Moenkopi Formation.

101.7 Beehivelike hills to the northwest and southeast are in the upper part of the Permian Cedar Mesa Sandstone. Hills to the northwest are capped by a thin Moenkopi and thin Organ Rock Shale. Halgaito redbeds form the broad slope down to the top of the Pennsylvanian Honaker Trail Formation which holds up the lower platform.

103.6 Excellent exposures of Cedar Mesa Sandstone forms beehives above well-bedded Halgaito Shale. This cliff is carved in the same massive sandstone which forms the buttes and columns in Monument Valley, to the northeast.

108.1 Arizona-Utah Boundary, in the western part of Monument Valley. The Mittens, to the east, and Mitchel Butte and Mesa in front of them to the southeast, have vertical cliffs of Cedar Mesa Sandstone with reddish Organ Rock and Moenkopi beds above and the reddish Halgaito Shale below. The highway is still at the top of the Honaker Trail Formation.

108.5 Junction of Side Roads East to Navajo Council Park and West to The Gap, Hospital, and Gouldings Trading Post. The road to the east leads 1.9 miles to a tribal campground, exhibit area, and overlook into Gypsum Creek and some of the spectacular erosional features of Monument Valley (fig. 10.13), as well as to the head of an access road which leads down into the scenic area. The road west leads 2.5 miles west to the hospital and Gouldings Trading Post in The Gap.

Beyond the junction U.S. Highway 163 crosses the valley of Mitchell Butte Wash mainly on recent windblown sediments.

Figure 10.13. View eastward into Monument Valley from the Navajo Council Park which is accessible by the side road east from Mile 108.5. The Mittens are in Cedar Mesa Sandstone, above a slope on Halgaito Shale. The lower part of the valley is on the top of the Honaker Trail Formation.

114.4 Excellent exposures of Halgaito redbeds occur in canyons and around the base of Eagle Mesa, to the west, Sentinel Mesa to the south, and Eagle Rock to the east (fig. 10.14) and along the road in Monument Pass. Massive Cedar Mesa Sandstone forms the cliffs capped by Organ Rock Shale or Hoskinnini Sandstone (fig. 10.15). Be-

Figure 10.14. Eagle Rock, east of the highway at Monument Pass, is capped by Hoskinnini Sandstone and Organ Rock Sahle, above vertical cliffs of massive Cedar Mesa Sandstone. Halgaito beds are the ledge and slope zone below, down to the top of the Honaker Trail Formation which is slightly below road level here.

yond the pass the highway drops through Halgaito beds onto the top of the fossiliferous Honaker Trail Formation toward the northeast (Fig. 10.16).

117.6 View Area to the south looks across Eagle Rock or Hulkito Wash to the Mittens area of Monument Valley (fig. 10-16).

121.0 Lower part of the Halgaito Formation and top of the Honaker Trail Formation at road level. The top of the Honaker Trail Formation forms the stripped surface over the Halgaito anticline between here and Mexican Hat.

122.2 Airport side road. Thin beds of the Honaker Trail Formation are exposed here at about the crest of the anticline over which we've been driving. Some of the lower limestone beds exposed in the gorges are fossiliferous.

126.1 Beginning of steep dip down the east flank of the Halgaito Anticline in outcrops of resistant limestone and softer gray shale of the top of the Honaker Trail Formation.

126.6 Scenic turnout on the north. On the skyline is Alhambra Rock (fig. 10.17), a small dark igneous intrusion, which is probably related to Agathle's Needle and similar intrusions to the south. Hogbacks and flatirons are well displayed on the west side of Raplee anticline to the east. These red and gray striped beds on the crest and flank of the structure are the same beds through which we are now riding. The brick red unit between here and the Raplee Anticline, in the vicinity of the Uranium Mill and Mexican Hat, is Halgaito Shale in the Mexican Hat Syncline. The curve of the highway, at about the base of the steep hill, is approximately at the contact of Honaker Trail and Halgaito Formations.

128.2 Side road east to Atlas Uranium Processing Plant at Mexican Hat. The mill was built here to handle ores mined in the White Canyon area and other areas nearby.

130.0 Cross San Juan River at Mexican Hat. The river marks the northern boundary of the Navajo Indian Reservation here. To the west down the gorge the tan Honaker Trail Formation is exposed below the more maroon and brick red Halgaito Shale that forms the upper red rim at the bridge.

131.2 Northeast end of Mexican Hat. U.S. Highway 163 continues northward, west of Raplee Anticline (fig. 10.18), in the contact zone of the Halgaito Shale on Honaker Trail Formation. Mexican Hat (fig. 10.19) is on the skyline to the northeast, just beyond the small synclinal, reef-limited, Mexican Hat oilfield.

133.0 Side road junction to the east. Gravel road leads down to the shore of the San Juan River and around the northern and eastern side of Mexican Hat. The highway is constructed on top of the uppermost Honaker Trail Limestone which is locally fossiliferous. Double road cuts are in Halgaito beds which contrast sharply with the underlying gray limestone.

134.5 Junction of Utah State Highway

SOUTHEASTERN UTAH
MONUMENT VALLEY - COMB RIDGE - BLUFF - ELK RIDGE - MONTICELLO

System	Series	Group	Formation	Thickness	Remarks
PERMIAN	Leon		White Rim Ss	0-20	thins northward from Monument Valley
			De Chelly Ss	0-550	
	Wolfcampian	Cutler Group	Organ Rock Shale	100—900	Organ Rock is maximum in Bluff subsurface
			Cedar Mesa Sandstone	500—1200	maximum in Elk Ridge area
			Halgaito Fm	400—500	mostly reddish brown siltstone and ss, some ls
PENNSYLVANIAN	Missourian	Hermosa Group	Honaker Trail Fm	500—1200	"Rico" of older reports. Hermosa Group thickens north-eastward toward Uncompahgre
	Desmoinsian		Paradox Fm — Oil Zones: Ismay, Deseret Creek, Akah, Barker Creek	100—2500	*Paradox Fm conodonts: Stone* (1968). Paradox salt. Aneth area oil production largely from carbonate reef complexes
	Atok		Pinkerton Trail Fm	100—500	
			Molas Fm	0-150	pre-Pennsylvanian not exposed but known from well data
MISS			Redwall Ls	300—650	thickens north-westward
DEV			Ouray Ls	50-140	
			Elbert Fm	160—350	
			McCracken Ss M	0-120	
			Aneth Fm	0-200	*fish plates*
CAMBRIAN			"Muav" Ls	400—600	
			Bright Angel Sh	90-300	
			Ignacio Qtzt (Tapeats)	200—300	
P Є			Precambrian Complex	—	

System	Formation	Member	Thickness	Remarks
OLIG	Abajo Mtn diorite		intrusive	28 M.Y. K-Ar (Armstrong, 1969)
CRET	Mancos Shale		0-400	*Gryphaea*
	Dakota Ss		80-150	*coal, plant fossils*
	Burro Canyon Fm		50-180	
JURASSIC	Morrison Fm	Brushy Basin Member	200—440	varicolored mudstone
		Westwater Can M	0-180	ss, mudstone reddish siltstone
		Recapture M	0-285	
		Salt Wash Ss Member	50—500	ss siltstone; U-V mines
	Bluff Sandstone		0-300	Permo-Jurassic wind and current directions: Poole (1961-62)
	Summerville Fm		100-200	
	Entrada Ss		100-150	
		Dewey Bridge M	20-70	
	Carmel Fm		0-120	present from Comb Ridge westward
	Navajo Ss		300—800	thickens westward
TRIASSIC	Kayenta Fm		50-250	
	Wingate Ss		250—350	
	Chinle Fm	Church Rock M	0-400	light brown sandy siltstone
		Owl Rock M	150-250	pale red siltstone and 10% ls
		Petrified Forest M	0-200	red purple, green bentonitic clay
		Moss Back M	0-250	
		Monitor Butte M	0-200	
		Shinarump M	0-200	
	Moenkopi Fm	upper memb	50-350	white or red ss
		Hoskinnini M	0-120	

Figure 10.15. Stratigraphic sections of rocks in southeastern Utah, in the Monument Valley-Monticello Area (from Hintze, 1973).

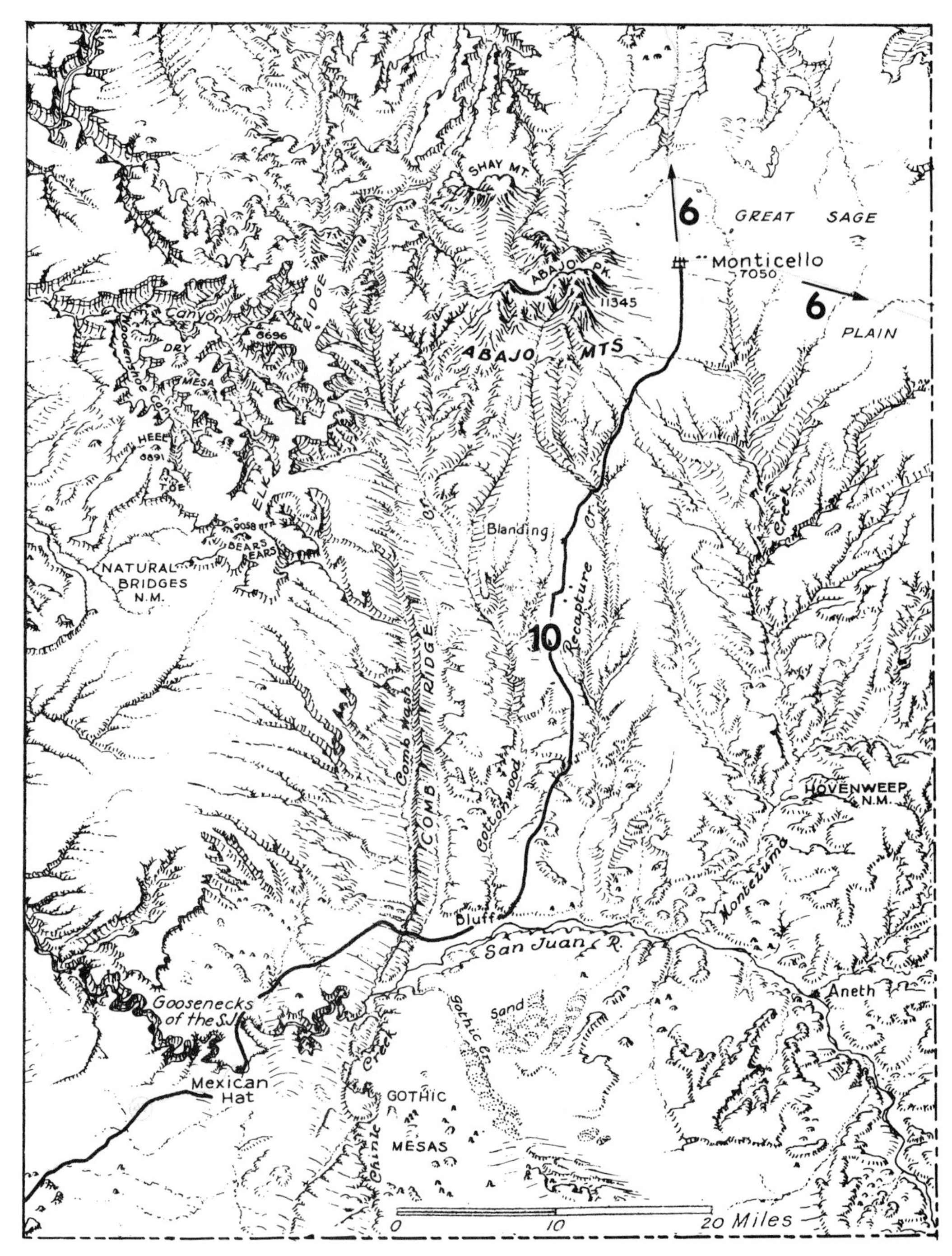

Figure 10.16. Index map of the northeastern part of Route 10 from Monument Valley to Monticello where Route 10 joins with Route 6.

Figure 10.17. Alhambra Rock or The Mule's Ear Diatreme, a small intrusion rises above gently dipping basal Halgaito Shale and upper Honaker Trail Formation, as seen northward from Mile 126.6 on the road log.

Figure 10.18. View northeastward across the Mexican Hat oil field toward the Raplee Anticline, from the northeast edge of the community of Mexican Hat at Mile 131.2. Well-bedded units in the Raplee Anticline are in the Honaker Trail Formation, but rounded hills immediately beyond the gravel quarry are Halgaito beds in the Mexican Hat Syncline. The Mexican Hat oil field is in the trough of the syncline. The petroleum accumulation is in a small shallowly buried reef in the Pennsylvanian sequence.

61 with U.S. Highway 163. State Highway 61 leads west toward Goosenecks of the San Juan River State Reserve and toward Natural Bridges National Monument. For a guide to this segment of the route to the Goosenecks see Geologic Guide Segment 10A. Continue ahead on U.S. Highway 163 toward Bluff and Blanding.

Figure 10.19. Mexican Hat as seen from the southeast from a graveled road that leads eastward from Mile 133.0. The Hat is a resistant unit in the Halgaito Formation.

Road cuts immediately beyond the junction are in marine limestone at the top of the Honaker Trail Formation which is dipping eastward into the Mexican Hat Syncline. The San Juan River has cut across the Raplee structure to the east and exposed the Paradox Formation in the gorge. West of the anticline, however, the river had adjusted to the softer Halgaito beds along the syncline.

135.1 Double road cuts through lower beds of Halgaito Shale. Honaker Trail beds are exposed in some of the deeper gullies.

137.7 Cross Lime Creek in Halgaito redbeds with gray Honaker Trail beds exposed to the east. Halgaito beds form hoodoos in the immediate vicinity and are capped by a thin Cedar Mesa Sandstone which forms the "mushrooms" on top of the cliffs.

138.8 Junction of side road to the Valley of the Gods scenic area (fig. 10.20). The

Figure 10.20. Monolithic "Gods" in the Valley of the Gods to the west of the highway. Rocks in the foreground are uppermost limestones of the Honaker Trail Formation and are abruptly overlain by slope-forming red Halgaito beds. The sheer cliff wall, the erosional remnants for which the area are noted, is in Cedar Mesa Sandstone. Valley of the Gods is accessable by a relatively primitive road that leads westward from Mile 138.8.

road to the west leads down across Lime Creek and around on the platform at the top of Honaker Trail beds into Valley of the Gods. The road leads 14.5 miles through the scenic area and connects with Utah State Highway 61 near Lees Ranch, northwest of Goosenecks of the San Juan turnoff. It is a moderately maintained primitive road, but suitable for most passenger cars with caution. Erosional remnants of the Cedar Mesa Sandstone have brick red Halgaito Shale bases and make up the monuments in Valley of the Gods.

140.7 Deep double road cuts through alternating resistant sandstone and easily eroded shale of the Halgaito-Honaker Trail transition beds. Fresh road cut exposures show the lenticular nature of some of the sandstone beds.

143.3 West flank of Limestone Ridge Anticline near the contact of Halgaito and Honaker Trail Formations.

144.5 Crest of the Limestone Ridge Anticline. To the east ledges of Honaker Trail Formation show well in gullies and start to dip eastward into Comb Wash.

146.0 Double road cuts through east-dipping Honaker Trail Formation with slope of the road and dip about the same. The Cedar Mesa Sandstone which forms the cliffs to the west thins to the east and grades into gypsiferous redbed facies on the eastern side of Lime Ridge (fig. 10.21).

Figure 10.21 Northeastward from approximately Mile 146 of the strong flexure down the east side of the Comb Wash Monocline. Halgaito beds, in the foreground near the road, are overlain by a single, thin tongue of Cedar Mesa Sandstone in the middle distance. Cliffs on the skyline are Wingate, Kayenta, and Navajo beds in the Comb Wash Monocline.

146.7 Cross through double road cuts in Halgaito redbeds.

147.3 Interbedded tan and redbeds in transition from Halgaito Formation into gypsiferous evaporite facies of Cedar Mesa Formation (fig. 10.21). This part of the section is capped by a thin sandstone 10 to 12 feet thick which is about all that is left of the massive cross-bedded Cedar Mesa Sandstone visible to the west. East of where the highway crosses through the light gray sandstone the gypsiferous upper part of the eastern Cedar Mesa evaporite facies is exposed in road cuts and along the anticlinal flank.

147.5 Massive gypsum in Cedar Mesa evaporite facies at top of series of road cuts. A short distance to the east a single, thin, light tan sandstone at the east end of the road cuts is the top of the Cedar Mesa sequence and marks the base of the Moenkopi Formation.

147.7 Moenkopi beds well exposed as reddish brown, thin, laminated, easily eroded sandstone and shale, particularly at the west bridge abutment.

148.2 Cross Comb Wash. The road climbs from Comb Wash floodplain and into debris-covered Chinle beds. Shinarump Conglomerate is exposed to the south as the tan or light gray resistant unit a short distance downstream from the bridge. Light grayish green and purple Chinle Formation is exposed down the wash beyond and in cuts along the west part of Comb Ridge.

148.5 Base of the Wingate Sandstone and top of Chinle Formation in road cuts along the west side of Comb Ridge with Wingate Sandstone in the deep double road cuts through the ridge. Comb Wash to the west is a subsequent stream in the easily eroded Triassic Moenkopi and Chinle rocks. Flatirons of upper Honaker Trail and lower Halgaito beds show on the east flank of Lime Ridge Anticline to the southwest.

148.8 Flaggy-bedded rocks near the east end of the deep road cut are in Kayenta Formation and cuts in the crest of the ridge to the east are in Navajo Sandstone. East of the east end of the cuts the slickrock country is on the top of the spectacularly cross-bedded Navajo Sandstone.

149.5 Cross-bedded Navajo Sandstone particularly well exposed on the north, approximately 100 yards west of an historical monument marking where men of the Hole-in-the-rock company spent a night in a snowstorm while trying to locate a trail across Comb Ridge in the later 1800s. Beyond the monument U.S. Highway 163 dips down and crosses Butler Wash, still in Navajo Sandstone.

150.2 East end of Navajo Sandstone road cuts, east of Butler Wash. Reddish Carmel Formation overlies the very light gray or tan Navajo Sandstone. To the north the high bluff above the shelf carved on the Carmel and Summerville Formation is formed by the almost Navajolike Bluff Sandstone Member of the Entrada Formation. Bluff Sandstone forms the cliffs across the San Juan River to the southeast and east of us as well.

151.6 The highway drops off a gravel-armoured high level terrace of the San Juan River onto a lower terrace. Navajo Sandstone is exposed in the river gorge to the southeast of us.

152.4 Junction side road to Sand Island Recreational Site. This is where boaters put into the river for trips down the San Juan River to Mexican Hat. East of the river the massive, white, Bluff Sandstone rises above the softer lower red Summerville Formation.

153.7 Drop from intermediate to lower gravel-veneered terraces here cut into reddish Carmel beds.

155.0 Bluff town limits at the southwest edge.

155.5 Cross Cottonwood Wash into the older part of Bluff. The stream is virtually choked with sediment and is consistently complexly braided and shallow.

156.4 Navajo Twins carved from Bluff Sandstone at the northwestern edge of town (fig. 10.22). Sunbonnet Rock, below, is a pedestal of lower reddish Summerville beds capped by a tumbled Bluff Sandstone block.

Figure 10.22. The Navajo Twins and Sunbonnet Rock at Mile 156, on the north edge of the community of Bluff. Upper light part of the Twins is in the Bluff Sandstone and the lower banded beds are Summerville Formation. Sunbonnet Rock is a tumbled block of Bluff Sandstone.

A short distance to the north, near St. Christopher's Mission Store, U.S. Highway 163 rises from the San Juan River Floodplain and climbs up through reddish Summerville Formation.

157.5 Upper narrows along highway where the route crosses through massive, white, cross-bedded Bluff Sandstone. Here the formation appears much like the older Navajo Sandstone and is probably an eolian deposit also. The road climbs northward onto the top of the Bluff Sandstone and into lower beds of the Morrison Formation.

160.6 Low hills to the east are in rocks of the lower Morrison Formation. Morrison beds are also well exposed in the tributaries to Cottonwood Creek to the west. Bright, light gray, Navajo Sandstone is exposed along the east side of Comb Ridge on farther to the west.

162.9 U.S. Highway 163 rises up and down through the Morrison section. Light colored resistant sandstone lenses here are remnants of old channel fillings. From the air these sandstone fillings stand out in relief as sinuous ridges for the softer enclosing floodplain deposits have been eroded away from around them.

164.5 Panoramic viewpoint, looking across Morrison beds. The San Juan Mountains, over 12,000 feet high, form the skyline in the far distance to the east, north of rounded Ute Mountain, elevation 9,900 feet. The Abajo Mountains are the promontories to the north and they have a high point with an elevation of 11,345 feet on the eastern peak. The top of Black Mesa, to the northwest, and White Mesa, to the north, are capped by Dakota Sandstone.

167.3 Junction of Utah State Highway 262 with U.S. Highway 163. State Highway 262 leads east to Hovenweep National Monument, Aneth Oil Field, and Cortez, Colorado. North of the junction Morrison beds show in the bluff (fig. 10.23) at the south edge of White Mesa as major channel-fill

Figure 10.23. Westward along the escarpment at the south end of White Mesa in the Morrison Formation as seen from approximately Mile 167.5. Dakota Sandstone caps the escarpment above the banded Morrison sequence.

sandstone lenses and softer mudstone. The mesa cap of Dakota Sandstone is prominent along the highway route.

168.1 Dakota Sandstone at the top of the climb to the flat uplands of White Mesa. The road on northward is across the Dakota-protected surface but gullies to the east and west show Morrison beds below.

172.5 Water tower for small community. The highway is still on Dakota Sandstone but here these Cretaceous rocks are blanketed by the reddish brown, loess-derived, Dove Creek soil. It is this soil which makes the area particularly suitable for agriculture.

180.5 Junction of Utah State Highway 95 with U.S. Highway 163. Utah State Highway 95 leads westward to Natural Bridges National Monument, Lake Powell in the Glen Canyon National Recreation Area, and Hanksville. Continue northward along U.S. Highway 163 to Blanding and Monticello. From here to Blanding the highway continues on Dove Creek soil over Dakota Sandstone.

181.4 Blanding airport to the west. The highway is still over Dove Creek soil and Dakota Formation, but rises toward the north over gravelly debris aprons from the southern part of the Abajo Mountains.

184.5 Blanding southern city limits. The community is built on the northern end of White Mesa, at the southeastern base of the Abajo Mountains.

185.5 Northern limits of Blanding, the road continues northward essentially on the top of the Dakota Sandstone which is veneered by Dove Creek soil and gravel debris.

189.5 Cross Recapture Creek, a deeply entrenched drainage from the southeastern part of the Abajo Mountains. Coarse-grained Dakota Sandstone is well exposed in cuts along the rim of the canyon and Morrison beds are exposed in the valley (fig. 10.24). The highway climbs toward the southeast side of the Abajo Mountains (fig. 10.25).

Figure 10.24. Northward across Recapture Creek towards the Abajo Mountains on the skyline. Rocks along either side of the canyon rim are Morrison beds, capped by relatively resistant Dakota Sandstone which forms the wooded uplands and the upper canyon rim. The Abajo Mountains are stocklike intrusions cutting rocks as young as Cretaceious.

195.2 Cross Devils Canyon, a tributary to Montezuma Creek to the east. Dakota Sandstone is exposed at the rim (fig. 10.26).

Figure 10.25. Northwestward from Mile 190 to the Abajo Mountains across the apron of coarse gravel which veneers a pediment surface on Dakota and Mancos beds.

200.4 Cross through Verdure graben, a downdropped block, with Dakota Sandstone dropped down against Burro Canyon or Morrison beds. The canyon here is approximately 200 feet deep near the community of Verdure.

201.4 Junction Side Road to Montezuma Canyon with U.S. Highway 163. The side road leads southeastward into Montezuma Canyon where uranium was produced from the Salt Wash Member of the Morrison Formation. Montezuma Canyon is also well-known for ancient Indian ruins.

206.1 Site of abandoned Monticello uranium mill east of the highway at the south end of town. Dakota Sandstone is exposed along the bluffs near the mill.

206.7 Monticello. Junction of U.S. Highway 163 with U.S. Highway 666 in Downtown Area. For a description of the geology to the north along U.S. Highway 160 and eastward along U.S. Highway 666 see Geologic Guide Segment 6.

Figure 10.26. Cross-bedded resistant Dakota Sandstone across Devil Canyon, a tributary to Montezuma Creek. Burro Canyon beds are exposed in the narrow notch in the base of the canyon and to the east of the road.

Segment 10A

0.0 Junction of Utah State Highway 61 with U.S. Highway 163, turn west onto Utah State Highway 61 toward Goosenecks State Reserve. The state highway from here to the park overlook is constructed in uppermost beds of the Honaker Trail Formation. Cedar Mesa Sandstone forms the vertical cliffs around Cedar Mesa toward the northwest and the erosional remnants and scenic pillars and columns above red Halgaito Shale in Valley of the Gods to the north.

0.9 Junction of Access Road to Goosenecks with Utah State Highway 61. Turn west toward the state reserve. The state park access road continues in uppermost fossiliferous limestones of the Honaker Trail Formation. Bryozoans, platy crustose algae, foraminifera, and some brachiopods occur in the limestone and associated shale.

2.9 Outcrops of the top limestone in a sharp bend and swithback along the road as it drops down into a gully.

4.6 Goosenecks of the San Juan River Overlook. The rocks at the upper surface are Honaker Trail Formation, but older Pennsylvanian rocks are exposed in the gorge. The San Juan River has entrenched a deep canyon across the nose of the Halgaito Anticline and has retained a sweeping meander pattern (fig. 10.27) established across a plain at a higher elevation. The meanders are such that the river must now flow 5 or 6 miles to make approximately 1 mile headway toward the west. The canyon here is approximately 1,000 feet deep.

Turn around and return to U.S. Highway 163 if you are continuing northward toward Blanding and Bluff or if you are continuing southwestward to Mexican Hat and Monument Valley. State Highway 61 connects northward with the Blanding-Lake Powell-Hanksville highway near the entrance to Natural Bridges National Monument.

Figure 10.27. Goosenecks of the San Juan River as seen from the north. The San Juan River has maintained broad sweeping meanders as it cut down into Pennsylvanian rocks exposed in the gorge.

SELECTED REFERENCES

Baars, D.L.; Parker, J.S.; and Chronic, John. 1967. Revised stratigraphic nomenclature of Pennsylvanian system, Paradox Basin. *Amer. Assoc. Petrol. Geol. Bull.,* v. 51, no. 3.

Baer, J.L. 1969. Paleoecology of cyclic sediments of the lower Green River Formation, central Utah. *Brigham Young Univ. Geol. Studies,* v. 16, part 1.

Baker, A.A. 1933. Geology and oil possibilities of the Moab District, Grand and San Juan Countries, Utah. *U.S. Geol. Survey Bull.* 841.

Bissell, J.J. 1962. Pennsylvanian-Permian Oquirrh Basin of Utah. *Brigham Young Univ. Geol. Studies,* v. 9, part 1.

———. 1968. Bonneville—an ice-age lake. *Brigham Young Univ. Geol. Studies,* v. 15, part 4.

Cook, E.F. 1960. Geologic atlas of Utah, Washington County. *Utah Geol. Mineral Survey Bull.* 70.

Gilbert, G.K. 1890. Lake Bonneville. *U.S. Geol. Survey Monograph 1.*

Gilluly, James. 1929. Geology and oil and gas prospects of part of the San Rafael Swell, Utah. *U.S. Geol. Survey Bull.* 806-C.

Hamblin, W.K. and Best, M.G. 1970. The western Grand Canyon District. *Utah Geol. Soc. Guidebook* 23.

Hardy, C.T. 1952. Eastern Sevier Valley, Sevier and Sanpete countries, Utah. *Utah Geol. Mineral. Survey Bull.* 43.

Hintze, L.F. 1962. Structure of the southern Wasatch Mountains and vicinity, Utah. *Brigham Young Univ. Geol. Studies,* v. 9, part 1.

———. 1973. Geologic road logs of western Utah and eastern Nevada. *Brigham Young Univ. Geol. Studies,* v. 20, part 2.

Hunt, C.B. 1956. Cenozoic geology of the Colorado Plateau. *U.S. Geol. Survey Prof.* Paper 279.

———. 1967. *Physiograph of the United States.* W.H. Freeman and Company.

King, P.B. 1959. *The Evolution of North America.* Princeton University Press.

———. 1969. Tectonics of North America—a discussion to accompany the tectonic map of North America. *U.S. Geol. Survey Prof. Paper* 628.

Melton, R.A. 1972. Paleoecology and paleoenvironments of the upper Honaker Trail Formation near Moab, Utah. *Brigham Young Univ. Geol. Studies,* v. 19, part 2.

Petersen, M.S.; Rigby, J.K.; and Hintze, L.F. 1973. *Historical Geology of North America.* Wm. C. Brown Company Publishers.

Peterson, J. A. 1972. Jurassic system. *Rocky Mountain Assoc. Geol. Atlas of the Rocky Mountain Region.*

Rigby, J.K. 1968. Guide to the geology and scenery of Spanish Fork Canyon along U.S. Highways 50 and 6 through the Southern Wasatch Mountains, Utah. *Brigham Young Univ. Geol. Studies,* v. 15, part 3.

———. Hamblin, W.K.; Matheny, R.; and Welsh, S.L. 1971. Guidebook to the Colorado River, Part 3: Moab to Hite, Utah through Canyonlands National Park. *Brigham Young Univ. Geol. Studies,* v. 18, part 2.

———. Hintze, L.F. and Welsh, S.L. 1974.

Geologic guide to the northwestern Colorado Plateau. *Brigham Young Univ. Geol. Studies,* v. 21, part 2.

Stokes, W.L. 1944. Morrison Formation and related deposits in and adjacent to the Colorado Plateau. *Geol. Soc. Amer. Bull.,* v. 55, no. 8.

———. 1952. Lower Cretaceous in Colorado Plateau: Amer. Assoc. Petrol. Geol. Bull., v. 36, no. 9.

Thornbury, W.D. 1965. *Regional Geomorphology of the United States.* John Wiley & Sons, Inc.

Tuttle, S.D. 1975. *Landforms and Landscapes, 2d ed.* Wm. C. Brown Company Publishers.

Wengerd, S.A. 1962. Pennsylvanian sedimentation in Paradox Basin, Four Corners Region. *Amer. Assoc. Petrol. Geol. Symposium, Penn. System in U.S.*

GLOSSARY

Aa—A Hawaiian term for basaltic lava flows typified by a rough, jagged, spinose, clinkery surface.

Abrasion—The mechanical wearing of solid materials by impact and friction.

Agglomerate—A fragmental volcanic rock consisting of large, somewhat rounded stones in a finer matrix, much like conglomerate in appearance but wholly volcanic in constitution.

Alluvium—Unconsolidated gravel, sand, and finer rock debris deposited principally by running water; adjective *alluvial.*

Angular unconformity—An arrangement in which older deformed stratified rocks have been truncated by erosion and younger layers have been laid down upon them with a different angle of inclination.

Antecedent stream—One which maintained its course in spite of localized uplift across its path; the stream anteceded the structure.

Anticlinal core—The mass of older rock in the heart of an anticline.

Anticlinal nose—The place where beds at the axis of a plunging anticline pass beneath the ground surface.

Anticline—A fold in stratified rock convex upward. Beds on the flanks are inclined outward.

Arkose—A clastic rock of granular texture, formed principally of large grains of quartz and feldspar.

Arroyo—The wide, flat-floored channel of an intermittent stream in dry country.

Ash—See volcanic ash.

Axis—The central line of an elongated geological structure such as an anticline or syncline.

Badlands—A region nearly devoid of vegetation where erosion, instead of carving hills and valleys of the ordinary type, has cut the land into an intricate maze of narrow ravines and sharp crests and pinnacles.

Barchan—An isolated, crescent-shaped dune, convex upwind.

Barranca—A vertical walled gully cut by an intermittent stream in relatively unconsolidated material.

Barrier island—Similar to a barrier beach but consisting of multiple instead of single ridges and commonly having dunes, vegetated zones, and swampy terraces extending lagoonward from the beach.

Basalt—A fine-grained black lava relatively rich in calcium, iron, and magnesium. The extrusive equivalent (in composition) of gabbro.

Basement—Old crystalline rocks upon which younger rocks have been deposited.

Batholith—A very large igneous body intruded into the earth's crust at considerable depth where it cooled slowly to form coarsely crystalline rock.

Bedding—The layered structure of sedimentary rocks.

Bedrock—Consolidated rock material of any sort.

Bench—A level or gently sloping area interrupting an otherwise steep slope.

Bench mark—An established mark, the elevation of which is accurately determined with respect to sea level.

Braided stream—A stream flowing in several dividing and reuniting channels resembling the strands of a braid, the cause of division being the obstruction by sediment deposited by the stream.

Breccia—A rock containing abundant angular fragments of rocks or minerals. These are sedimentary breccias, volcanic breccias, tectonic breccias, landslide breccias, and other types.

Butte—Detached and commonly flat-topped hills and ridges which rise abruptly, and reach too high to be called hills or ridges, and not high enough to be called mountains.

Calcareous—Rich in calcite.

Calcite—A common mineral composed of calcium, carbon, and oxygen ($CaCO_3$). The principal source of cement.

Caldera—A large basin-shaped volcanic depression, more or less circular or craterlike in form, the diameter of which is many times greater than that of the included volcanic vent or vents.

Caliche—A calcareous deposit formed within dry-region soils by weathering.

Capture—See stream capture.

Carbonate rocks—Those composed of the minerals calcite (calcium carbonate) and dolomite (calcium-magnesium carbonate).

Chert—Cryptocrystalline varieties of silica regardless of color.

Cinder cone—A conical elevation formed by accumulation of volcanic ash or clinkerlike material around a vent.

Cleavage—The facility to break along parallel smooth planes, especially in minerals, but also in rocks.

Concretion—A nodular or irregular concentration of certain authigenic constituents of sedimentary rocks and tuffs; developed by the localized deposition of material from solution, generally about a central nucleus.

Conglomerate—A sedimentary rock consisting of larger rounded rock and mineral fragments embedded in a finer, usually sandy matrix and all cemented together.

Consequent stream—One which follows a course that is a direct consequence of the original slope of the surface on which it developed.

Cross bedded—The arrangement of laminations of strata transverse or oblique to the main planes of stratification of the strata concerned; inclined, often lenticular, beds between the main bedding planes.

Crystal—A regular, solid, geometrical form bounded by plane surfaces expressing an internal ordered arrangement of atoms.

Crystalline—Substances having fixed internal atomic arrangements.

Crystalline rocks—A term commonly applied to mixed igneous and metamorphic rocks, or to either separately.

Cuesta—A ridge or escarpment held up by a resistant bed that has a perceptibly but gently sloping attitude.

Debris—Broken-up and usually partly decomposed rock materials.

Debris cone—A cone-shaped accumulation of rock debris at the mouth of a gully or small canyon, usually smaller, steeper, and often rougher than an alluvial fan.

Debris flow—A flow of usually wet, muddy rock debris of mixed sizes, much like a slurry of freshly mixed concrete pouring down a chute.

Decomposition—The chemical breakdown of rocks and minerals.

Delta—An alluvial deposit, usually triangular, at the mouth of a river.

Desert pavement—An armor of closely fitted stones, one layer thick, on the surface of alluvial material. Basically a residual accumulation of larger fragments owing to removal of fine particles.

Desert varnish—A thin coating rich in iron and manganese on rock surfaces developed by weathering.

Diatomite—A sedimentary rock consisting almost entirely of the siliceous skeletons of single-celled algae.

Differential erosion—The more rapid erosion of one portion of the earth's surface as compared with another.

Dike—A sheetlike body of igneous rock formed by intrusion along a fracture.

Diorite—A coarse-grained intrusive igneous rock about midway between a granite and a gabbro in chemical and mineralogical composition.

Dip—The direction and degree of inclination (from horizontal) of a sedimentary bed or any other geological planar feature.

Disconformity—An unconformity or erosional surface in which the beds on opposite sides are parallel; unconformity between parallel strata.

Disintegration—The physical breakup of rocks and minerals.

Dolomite—A sedimentary rock composed of the mineral dolomite, a calcium-magnesium carbonate.

Dome—A topographic dome is a roughly circular, upwardly convex land form. A structural dome in sedimentary rocks involves an outward dip or inclination of the beds in all directions. A volcanic dome is a domelike extrusion of highly viscous lava.

Earth flow—A form of mass movement in which relatively unconsolidated surface material, usually weathered, flows down a hillside.

Embayment—An indentation along a shoreline, mountain front, or any other natural linear feature.

End moraine—A moraine deposited at the lower end of an ice stream or outer end of an ice lobe.

Entrenched stream—A stream in a narrow, meandering trench well beneath the general surface of the adjacent upland.

Eolian—Consisting of materials drifted and arranged by the wind.

Epicenter—The spot on the earth's surface directly above the subsurface point at which an earthquake shock originates.

Erosion—The removal of rock material by any natural process.

Evaporites—Sediments deposited from aqueous solution as a result of extensive or total evaporation of the solvent.

Extrusive rock—Rock extruded onto the earth's surface, usually in molten condition (lava).

Faceted spur—The end of a ridge which has been truncated or steeply beveled by stream erosion, glaciation, or faulting.

Fan—A deposit, usually alluvial, of rock debris at the foot of a steep slope (mountain face) with an apex at the mountain base (canyon mouth) and a radial, fanlike, divergence therefrom.

Fanglomerate—The consolidated deposits of an alluvial fan; a variety of conglomerate which is coarse, ill-sorted, and contains angular stones.

Fault—A fracture along which blocks of the earth's crust have slipped past each other.

Fault ridge—An elevated, elongate block lying between two essentially parallel faults.

Fault slice—A narrow segment of rock caught between two essentially parallel, closely adjacent faults.

Fault zone—A zone in the earth's crust consisting of many roughly parallel, overlapping, closely spaced faults and fractures, may be up to several miles wide.

Feldspar—An abundant rock forming class of minerals composed of aluminum, silicon, oxygen, and one or more of the alkalies, sodium, calcium, and potassium.

Flatiron ridge—A linear ridge with one very smooth flank formed by erosion of tilted sedimentary rocks and given a triangular shape by cross cutting canyons.

Floodplain—That portion of a river valley, adjacent to the river channel, which is built of sediments and is covered with water when the river overflows its banks at flood stage.

Fluvial—Features of erosion or deposition created by running water.

Foliation—A crude banding formed in rocks by metamorphism, less regular than the bedding of sedimentary rocks.

Formation—A geological formation is a rock unit of distinctive characteristics which formed over a limited span of time and under some uniformity of conditions. To a geologist it is a rock body of some considerable areal extent which can be recognized, named, and mapped.

Gabbro—A dark, coarse-grained intrusive igneous rock richer in iron, magnesium, and calcium and poorer in silica than granite.

Geophysical exploration—Subsurface exploration of rocks and structures carried on by indirect means such as gravity or magnetic variations.

Geothermal—Involving heat from within the earth.

Geyser—A type of intermittent spring in which the discharge is caused at more or less regular and frequent intervals by the expansive force of highly heated steam.

Gneiss—A coarse-grained metamorphic rock with irregular banding (foliation).

Gorge—A narrow, steep-walled passage cut into rock by a stream.

Graben—A sizeable block of the earth's crust dropped down between two faults steeply inclined inward, giving a keystone shape to the block, longer than it is wide.

Grain—Used here for a perceptible linear pattern in landscape features of a region, usually reflecting a similar pattern in underlying rock structure.

Granite—A common, coarse-grained, igneous intrusive rock relatively rich in silica, potassium, and sodium.

Granitic—A term commonly used for many coarse-grained igneous intrusive rocks not strictly of granite composition.

Granodiorite—A coarse-grained, igneous intrusive rock halfway between a granite and a diorite on the scale of rock composition.

Gully—A small ravine cut by running water.

Gypsum—Alabaster. Selenite, satin spar. A common mineral of evaporites.

Hanging valley—A tributary valley the floor of which is much higher at its mouth than the floor of the trunk valley.

Hogback—A ridge composed of a resistant layer within steeply tilted eroded strata.

Hoodoo—Pillars developed by erosion of horizontal strata of varying hardness in regions where most rainfall is concentrated during a short period of the year.

Hummock—A mound or knoll; a small elevation; hillock.

Igneous rocks—A class of rocks formed by crystallization from a molten state.

Inclusion—A fragment of older rock inclosed (included) within an igneous rock.

Incompetent—A rock which is relatively weak and responds readily to pressure by crumpling or by flow.

Intermittent stream—One which does not have a continuous or perennial flow.

Intrusive—Rocks or rock masses which have been intruded or injected into other rock, usually in a molten state.

Joint—Fracture in rock, generally more or less vertical or transverse to bedding, along which no appreciable movement has occurred.

Laccolith—A concordant, intrusive body that has domed up the overlying rocks and also has a floor that is generally horizontal but may be convex downward.

Lacustrine—Produced by or belonging to lakes.

Landslide—Sudden movement of earth and rocks down a steep slope.

Lateral fault—One on which the displacement is sidewise rather than up-down.

Lateral moraine—A ridgelike deposit of bouldery ill-sorted debris laid down along the lateral margin of a valley glacier.

Lava—The term is used both for molten rock material extruded onto the earth's surface and for the consolidated (crystallized) rock.

Left lateral fault—One on which the opposing block appears to have moved to the left, no matter which side you stand on.

Limb—One of the two sides of an anticline or syncline.

Limestone—A sedimentary rock composed wholly or almost wholly of the mineral calcite.

Magma—Molten rock within the earth's crust.

Marble—Recrystallized limestone or dolomite; a metamorphic rock.

Marine—The ocean environment; marine sediments are those deposited in the ocean.

Mass movements—The movement, usually down slope, of a mass of rock or rock debris by gravity, not transported by some other agent such as ice or water.

Matrix—The fine-grained constituents of a rock in which coarser particles are embedded.

Mesa—A flat-topped tableland with steep sides.

Mesozoic—One of the eras of the geological time scale extending from 70 to 230 m.y. ago.

Metamorphic rocks—Those which have undergone such marked physical change because of heat or pressure or both as to be distinct from the original rock. The process is *metamorphism.*

Metavolcanic—Rocks formed by metamorphism of volcanic materials.

m.y.—An abbreviation for a million years.

Mineral—A homogeneous, naturally occurring, solid substance of inorganic composition, consistent physical properties, and specified chemical composition.

Monocline—A fold with strata dipping in one direction, between segments with beds that are nearly flat.

Monolithologic breccia—A breccia formed of fragments of only one kind of rock.

Monomineralic rock—One composed of only one mineral; for example, limestone and dolomite.

Moraine—A deposit of coarse, ill-sorted rock debris laid down by glacial ice without intervention of any other agent, such as running water.

Mudcrack—Desiccation crack.

Mudflow—A form of mass movement involving the flow of mud, usually containing coarser rock debris, in which instance the term debris flow is equally applicable.

Mud pot—A shallow, hot-spring pit filled with bubbling mud.

Mudstone—A fine-grained sedimentary rock which is hard to characterize as shale or siltstone because of massiveness or poor sorting.

Normal fault—A fault in which the hanging wall has more downward, relative to the footwall.

Nose—*See* anticlinal nose.

Oblique air photo—One taken with the axis of the camera tilted from vertical. If the horizon shows, it is a high-oblique photo.

Obsidian—Natural volcanic glass. Lava which cooled so rapidly that it didn't crystallize.

Odometer—An instrument for measuring distance.

Oolite—A spherical to ellipsoidal body, up to 2.00 mm in diameter, which has concentric or radial structure or both.

Ore deposit—An accumulation of metallic minerals that can be mined at a profit. The minerals are termed *ore minerals,* and the aggregate is termed *ore.*

Outcrop—An exposure of bedrock at the surface.

Outlier—Those detached portions of strata which stand out from the main body, and appear to have been separated by the denuding force of water.

Pahoehoe—A Hawaiian term for basaltic lava flows typified by a smooth, billowy, or ropy surface.

Paleozoic—A major era of the geological time scale embracing the interval from 230 to 600 m.y.

Pediment—A relatively smooth, gently sloping surface produced by erosion at the foot of a steeper face, usually a mountain.

Pegmatite—A very coarse-grained igneous rock formed by the fluids given off in the late stage of crystallization of an igneous body; most often close to granite in composition.

Pendant—A large mass of metamorphic rock within a younger intrusive rock, thought to have hung down into the original intrusive body from the roof of the intrusive chamber.

Perched water table—If the underlying bed is of small extent but impervious it will force water contained in overlying porous material to the surface. In many places such water lies far above the ordinary water table and constitutes what is called a perched water table.

Pisolite—A spherical or subspherical, accretionary body over 2 mm in diameter.

Placer deposit—A water laid accumulation of rock debris containing a concentration of heavy, physically and chemically resistant, valuable mineral such as diamond, gold, or platinum. Such minerals are described as *placer minerals.*

Playa—The flat, smooth floor of a dry lake in desert regions.

Pleistocene—An epoch within the Cenozoic Era of the geological time scale. Usually taken to embrace the last 2 million years.

Plug—A small, cylindrical, near-surface, igneous intrusive body.

Plunge—The inclination from horizontal of the long axis of a fold or warp.

Pluvial period—An interval of cooler, wetter conditions in a dry region, coincident with a phase of glaciation in colder, better-watered areas.

Potassium-argon—A method of absolute dating of rocks and minerals using the ratio of radioactive potassium to its daughter product, the argon 40 isotope.

Pothole—A narrow cylindrical hole worn into solid rock by a fixed vortex in a stream.

Precambrian—All rocks older than Paleozoic.

Pumice—Frothy rock glass, so light that it floats.

Pyroclastic—Hot or firey (pyro) fragmental (clastic) debris thrown out of an explosive volcanic vent.

Pyroxene—A common igneous- and metamorphic-rock family of minerals, often green to black, and ranging widely in composition.

Quartz—One of our most common minerals, hard and chemically resistant, composed of silicon and oxygen (SiO_2).

Quartzite—A rock formed by metamorphism of sandstone, which is hard, coherent, and consists of quartz.

Radioactive—The property of some elements to

spontaneously change into other elements with the emission of charged particles, usually accompanied by generation of heat.

Radio-carbon—The radioactive isotope of carbon (^{14}C) which disintegrates at a known rate. It is used to determine geological ages up to about 40,000 years.

Rare earths—The oxide compounds of rare-earth elements, such as cerium, ytterbium, neodymium, and others.

Recharge well—A well designed for injection of fluids into the ground.

Relief—Topographic relief is the difference in elevation of contiguous parts of a landscape, valley to peak.

Rhyolite—An extrusive igneous rock of granitic composition, fine-grained, often light-colored to red.

Rift—As used here, refers to the shallow topographic trench, a mile or two wide, along the trace of a major fault.

Right-lateral fault—One on which the displacement of the opposing block appears to have been to the right, no matter on which side the observer stands.

Rillensteine—A stone with small, interlacing, wormlike solution channels on its surface. They form on soluble rocks, most commonly limestone.

Rock—An aggregate of minerals.

Rock cleavage—The facility to break along parallel smooth planes within a mass of rock.

Rockfall—The relatively free fall of rock masses from steep bedrock faces.

Rock glacier—An accumulation of large angular blocks of rock, usually lobate in form with steep margins, that moves slowly by creep.

Sagpond—A pond occupying a depression along the trace of a major fault, usually where a block within the zone has sunk.

Salt anticline or salt dome—An anticline in which the core is composed of rock salt. A salt anticline is elongate whereas a salt dome is essentially circular in plan.

Sandstone—A sedimentary rock formed by cementation of sand-size particles.

Scarp—A straight steep bank or face which can be a few feet to thousands of feet high, like the east face of the Sierra Nevada.

Schist—A finer-grained and more thinly and regularly foliated metamorphic rock than gneiss.

Scoria—Small fragments of porous volcanic rock thrown out of an explosive volcanic vent. Usually black or red and up to 1 1/2 inches in diameter.

Sedimentary rocks—A class of rocks of secondary origin, made up of transported and deposited rock and mineral particles and of chemical substances derived from weathering.

Septum—An older mass of metamorphic rock separating two adjacent intrusive igneous bodies.

Serpentinite—A rock consisting largely of the mineral serpentine, a hydrous magnesium silicate, produced by alteration of igneous rocks rich in iron and magnesium.

Shale—A sedimentary rock consisting largely of very fine mineral particles, laid down in thin layers.

Siliceous—Rich in silica, SiO_2.

Siltstone—A fine-grained, well-bedded sedimentary rock composed of silt, finer than sand and coarser than clay.

Slate—A weakly metamorphosed rock derived from shale by compaction with the development of closely spaced, smooth, (slaty cleavage).

Slump—The downward slipping of a mass of rock or unconsolidated material of any size, moving as a unit or as several subsidiary units, usually with backward rotation on a more or less horizontal axis parallel to the cliff or slope from which it descends.

Soapstone—A massive, soft, slippery rock composed of the mineral talc, a hydrous magnesium silicate.

Sorting—The arrangement of particles by size.

Spur—The subordinate ridges extending from the crest of a larger ridge.

Strata—Layers of a sedimentary rock. Bedded rocks are *stratified.*

Stream capture—The diversion of the headwaters of a stream owing to headward growth of an adjacent stream.

Strike valley—A valley eroded parallel to the strike of the underlying strata.

Structure—Phenomena that determine the geometrical relationships of rock units, such as folds, faults, and fractures.

Superimposed stream—One which has cut down through an overlying mantle into rocks of different character and structure.

Syenite—An intrusive igneous rock much like granite but lacking or very low in quartz.

Syncline—A down-fold in layered rocks which is

concave upward. Beds on the flanks are inclined inward.

Talus—Rock fragments that accumulate in a pile at the base of a ridge or cliff.

Terrace—A geometrical form consisting of a flat tread and a steep riser or cliff. Stream terraces, lake terraces, marine terraces, and structural terraces are distinguished in geology.

Terrestrial—Deposits laid down on land as contrasted to the sea; terrestrial conditions as compared to marine conditions.

Tertiary—A period of the Cenozoic Era embracing the time from 70 to 2 m.y. ago.

Thrust fault—A gently inclined fault along which one block is thrust over another.

Thrust plate—The upper block of a thrust fault.

Tidal flat—A marshy or muddy land area which is covered and uncovered by the rise and fall of the tide.

Till—Ill-sorted, mixed fine and coarse rock debris deposited directly from glacial ice.

Travertine—An accumulation of calcium carbonate formed by deposition from ground or surface waters, commonly porous and cellular.

Tuff—A fine-grained rock composed of volcanic ash.

Unconformity—A surface of erosion separating younger strata from older rocks.

Varnish—See desert varnish.

Vein—A sheetlike deposit of mineral matter along a fracture.

Ventifact—A stone whose shape and surface characteristics have been modified by natural sandblasting.

Vertical air photo—One taken with the axis of the camera pointed straight down toward the ground.

Volcanic ash—Fine-grained (less than 1/8 inch diameter) volcanic debris, often glassy, explosively erupted from a volcanic vent.

Volcanic cinders—Like volcanic ash but coarser, 1/8 to 1 inch. Fragments are highly porous.

Volcanic neck—The solidified lava filling the vent or neck of an ancient volcano. Exposed by erosion.

Volcanic tuff—A compacted deposit consisting of ash, cinders, and occasionally larger fragments of solid volcanic rock. If the latter are numerous, it is known as a tuff-breccia.

Warp—A part of the earth's crust which has been broadly bent.

Water gap—A gap in a ridge, cut and still occupied by the stream that cut it.

Water table—The level beneath the ground surface below which all openings in rocks are filled with water.

Wind gap—A gap or saddle in a ridge now abandoned by the stream that cut it.

Wineglass canyon—A canyon cut into the steep face of a mountain range. The fan at the mountain foot is the base, the gorge approaching the canyon mouth is the stem, and the headwaters basin is the bowl.

Zircon—A mineral found in small amounts in many igneous and metamorphic rock, a zirconium silicate and a gemstone. Chemically and mechanically tough.